Engineering Sustainable Development 2019

Seoul, South Korea
12 – 13 December 2019

ISBN: 978-1-7138-0581-6

Printed from e-media with permission by:

Curran Associates, Inc.
57 Morehouse Lane
Red Hook, NY 12571

Some format issues inherent in the e-media version may also appear in this print version.

Printed with permission by Curran Associates, Inc. (2020)

For permission requests, please contact AIChE
at the address below.

AIChE
120 Wall Street, FL 23
New York, NY 10005-4020

Phone: (800) 242-4363
Fax: (203) 775-5177

www.aiche.org

Additional copies of this publication are available from:

Curran Associates, Inc.
57 Morehouse Lane
Red Hook, NY 12571 USA
Phone: 845-758-0400
Fax: 845-758-2633
Email: curran@proceedings.com
Web: www.proceedings.com

TABLE OF CONTENTS

PLENARY SESSION II

INVITED SESSION 4 - SUSTAINABLE ENERGY AND WATER PRACTICES

PARALLEL SESSION 5A - SUSTAINABLE ENERGY

PARALLEL SESSION 5B - ECONOMIC, SOCIAL, CLIMATIC AND ENVIRONMENTAL ISSUES

Why Universities need to focus on Innovation and Sustainable Development

ENGINEERING
SUSTAINABLE
DEVELOPMENT

Conference: Technical and Engineering Challenges of Addressing Sustainable Development

Year: 2019

Proceeding: Engineering Sustainable Development 2019

Group: Abstract Submissions

Session: Plenary Session I

Time: Thursday, December 12, 2019 - 10:15am-10:40am

Authors:

Sch□nleber, C., Association of Pacific Rim Universities (APRU)

As a network of 51 leading universities the Association of Pacific Rim Universities (APRU) brings together leading research strength from 18 economies across Asia, Australasia and the Americas. The Association leverages the research capability of its members to collaborate with thought leaders, policy makers and external organizations to address key societal, environmental and economic challenges of the region. In her presentation Christina SchÃ¶nleber will outline why it is universitiesâ€™ responsibility to engage externally and collaboratively, acting across borders and regions to address existential global challenges. Drawing on APRUâ €™s experience of more than 20 years of working collectively with members and external stakeholders the presentation will address the importance of international cooperation to nurture future leaders of tomorrow and draw on latest research to develop solution to support policy makers for the greater common good of society and the world.

Invited Talk: SKI's Efforts on Climate Change and Waste Issue

Conference: Technical and Engineering Challenges of Addressing Sustainable Development

Year: 2019

Proceeding: Engineering Sustainable Development 2019

Group: Abstract Submissions

Session: Plenary Session I

Time: Thursday, December 12, 2019 - 10:40am-11:05am

Authors:

Lee, S. J., SK Innovation

SK Innovation(â€œSKIâ€) sets agenda for solving environmental issues such as climate change and challenges of waste recycling in pursuit of creating social value.

Since the industrialization, the level of CO_2 in the atmosphere has increased rapidly, and the current trend of emissions is expected to drive into an irreversible climate catastrophe at the end of 21st century. In addition to reducing CO_2 emissions directly, it is also essential to R&D new technologies such as CO_2 capture & utilization (CCU). SKI is investigating and evaluating CO_2 utilization technologies for determining their potentials in technical as well as economical aspects.

In addition to those efforts on mitigating climate change, developing solutions for plastic pollution which recently became one of the biggest environmental concerns is being focused. Due to attractive properties, lower costs and more, global production of virgin polymer reached up to 330 million tons in 2016 but only 16% of them have been collected for recycling and over 20% are presumed to be leaked into our nature. SKI performs long-term strategy for moving towards a circular economy combined with its Open Innovation efforts to accelerate the development of innovative recycling technologies.

Due to the exponential growth of battery electric vehicle, it is forecasted that ~150GWH of EV batteries will be retired in 2030, which might cause social problems. SKI has been developing a battery recycle technology to recover lithium as LiOH from waste batteries.

Plenary Talk: Development of a Simple Electrochemical Advanced Oxidation Process for Potable Reuse

ENGINEERING SUSTAINABLE DEVELOPMENT

Conference: Technical and Engineering Challenges of Addressing Sustainable Development

Year: 2019

Proceeding: Engineering Sustainable Development 2019

Group: Abstract Submissions

Session: Plenary Session I

Time: Thursday, December 12, 2019 - 11:05am-11:30am

Authors:

Mitch, W., Yale University

The production of hydroxyl radicals (*OH) by the UV photolysis of hydrogen peroxide (H_2O_2), an advanced oxidation process (AOP), is a key component of advanced treatment trains for the purification of municipal wastewater for potable reuse. However, due to the low absorption coefficient of H_2O_2 at 254 nm, the production of radicals is very inefficient, and only ~10% of the H_2O_2 is used despite the application of up to 1000 mJ/cm^2 UV fluence. This presentation describes the development of an electrochemical alternative pathway to *OH production. Previous electrochemical AOPs have focused on electro-Fenton processes, wherein Fe^{2+} provides an electron to H_2O_2 to produce *OH, and the resulting Fe^{3+} is converted back to Fe^{2+} at a cathode. This process is inefficient under the pH conditions of natural waters. Moreover, while it can function in batch reactors, the dissolved iron would be lost into the effluent of a flow-through reactor. We demonstrate the production of *OH by feeding electrons from the grid to H_2O_2 directly via a stainless steel cathode mesh. Potable reuse facilities typically must demonstrate 0.5-log removal of 1,4-dioxane to validate their AOP process. We demonstrate the achievement of this target with the electrochemical system over timescales relevant to full-scale treatment (i.e., minutes). We also demonstrate that this process requires lower H_2O_2 concentrations, and uses the H_2O_2 more efficiently.

Invited Talk: 2D and 3D Nanostructuring Strategies for Thermoelectric Materials

ENGINEERING
SUSTAINABLE
DEVELOPMENT

Conference: Technical and Engineering Challenges of Addressing Sustainable Development

Year: 2019

Proceeding: Engineering Sustainable Development 2019

Group: Abstract Submissions

Session: Invited Parallel Session 1A - Sustainable Energy

Time: Thursday, December 12, 2019 - 11:45am-12:10pm

Authors:

Jeon, S., KAIST

Thermoelectric materials have attracted increased research attention as the implementation of various nanostructures has potential to improve both their performance and applicability. A traditional limitation of thermoelectric performance in bulk materials is the interconnected nature of the individual parameters (for example, it is difficult to decrease thermal conductivity while maintaining electrical conductivity), but through the rational design of nanoscale structures, it is possible to decouple these relationships and greatly enhance the performance. For 2D strategies, newly investigated materials such as graphene, transition metal dichalcogenides, black phosphorus, etc. are attractive thanks to not only their unique thermoelectric properties, but also potential advantages in ease of processing, flexibility, and lack of rare or toxic constituent elements. For 3D strategies, the use of induced porosity, assembly of various nanostructures, and nanoscale lithography all offer specific advantages over bulk materials of the same chemical composition, most notably decreased thermal conductivity due to phonon scattering and enhanced Seebeck coefficient due to energy filtering. In this talk, my groupâ€™s techniques and strategies for 2D and 3D thermoelectric materials will be provided, along with how new 3D patterning methods and 2D materials can be utilized.

Invited Talk: Metal Halide Perovskite Optoelectronic Devices

ENGINEERING
SUSTAINABLE
DEVELOPMENT

Conference: Technical and Engineering Challenges of Addressing Sustainable Development

Year: 2019

Proceeding: Engineering Sustainable Development 2019

Group: Abstract Submissions

Session: Invited Parallel Session 1A - Sustainable Energy

Time: Thursday, December 12, 2019 - 12:10pm-12:35pm

Authors:

Im, S. H., Korea University

Solar energy has been considered as a clean, renewable, and sustainable energy and an electrical energy, which is the most safe and convenient energy form to live daily life, has been obtained from the direct and indirect products of it. Solar cell is a device that directly generates electricity from solar light energy so intensive studies have been done to develop highly efficient solar cells with low cost to expand its usage for power generation. Although crystalline Si solar cells have been successfully commercialized for power generation, their power generation cost is still more expensive than the fossil fuels due to expensive vacuum process. Hence, the development of cheap solution processable next-generation solar cells has been studied for several decades. Recently, metal halide perovskite has considered as a promising candidate for next-generation solar cell because of their excellent properties such as strong absorptivity, convenient bandgap tunability, small exciton binding energy and solution processability. Here, I would like to share research trends of metal halide perovskite solar cells and what we have studied.

Invited Talk: Integrated Biorefinery Strategies for Waste and Biomass Valorisation

ENGINEERING SUSTAINABLE DEVELOPMENT

Conference: Technical and Engineering Challenges of Addressing Sustainable Development

Year: 2019

Proceeding: Engineering Sustainable Development 2019

Group: Abstract Submissions

Session: Invited Parallel Session 1A - Sustainable Energy

Time: Thursday, December 12, 2019 - 12:35pm-1:00pm

Authors:

Lin, C. S. K., City University of Hong Kong

There is a growing demand to establish biotechnology-based processes for chemicals, materials and fuels production which can decrease our dependency on dwindling oil reserves and reduce greenhouse gas emissions. These novel bioprocesses are incorporated into biorefineries, which are integral units converting different renewable-derived feedstocks into a range of useful products as diverse as those from petroleum using environmentally friendly technologies [1-2]. A number of agricultural-based biorefinery processes have been developed and realised at commercial scale since the past decade. At the same time, there are growing concerns over organic waste generation and insufficient resource supply due to the ever-increasing global population and the cradle-to-grave system of material flows in the linear economy.

Therefore, the research in our team aims at valorisation of organic waste materials through bioconversion processes to recover their inherent nutrients for transformation into value-added products. Waste-based biorefinery not only provides a mean for waste treatment, but also promotes the development of a circular economy by valorisation of wastes into value-added products [3]. In this talk, we aim to provide an overview of recent efforts from our group in leading the future of global researchers. The first part of the talk will cover several projects which serve as examples to demonstrate the recent development of integrated biorefinery strategies for valorisation of food and textile wastes. Due to the different characteristics of the waste materials, the bio-processes were specifically designed and demonstrated in both laboratory and pilot-scales, accompanied with techno-economic assessment for evaluation of technical and economic feasibility.

Invited Talk: Sustainable Biofuel Production Using Renewable Electricity

ENGINEERING
SUSTAINABLE
DEVELOPMENT

Conference: Technical and Engineering Challenges of Addressing Sustainable Development

Year: 2019

Proceeding: Engineering Sustainable Development 2019

Group: Abstract Submissions

Session: Invited Parallel Session 1A - Sustainable Energy

Time: Thursday, December 12, 2019 - 1:00pm-1:25pm

Authors:

Lam, C. H., City University of Hong Kong

Biomass fast pyrolysis liquid (aka bio-oil) has been long regarded as one of the most promising renewable carbon feedstocks. However, freshly pyrolyzed bio-oil is acidic and corrosive containing large quantities of carbonyl and phenolic compounds that polymerize under acidic conditions. This presentation shares an electrochemical strategy to hydrogenate the carbonyl groups using only earth abundant metal catalysts. The upgraded biofuel goes through a series of hydrodeoxygenation under ambient conditions (60 ^{o}C and 1 atm). The product displays improved stability and higher specific energy. In addition to covering reaction mechanisms, this talk will also discuss why electrocatalysis is an important technique in the biofuel circle. Electrocatalysis enables direct installation of renewable energy into liquified biomass, which effectively creates a renewable energy hydrocarbon battery that is compatible with the fueling infrastructure.

Invited Talk: Design and Synthesis of Nanoporous Materials for Water Treatment Clean Technology

ENGINEERING
SUSTAINABLE
DEVELOPMENT

Conference: Technical and Engineering Challenges of Addressing Sustainable Development

Year: 2019

Proceeding: Engineering Sustainable Development 2019

Group: Abstract Submissions

Session: Invited Parallel Session 1B - Water Sustainability and Security

Time: Thursday, December 12, 2019 - 11:45am-12:10pm

Authors:

Kim, J., Kookmin University
Yamauchi, Y., The University of Queensland

Recently, nanoporous functional materials such as carbons, metal oxides, metal-organic frameworks, and hybrid composites have received significant interest in materials science because of their advantages of controllable porosity, large surface area, good thermal/chemical stability, high electrical conductivity, catalytic activity, easy modification with other elements and materials, etc. Thus, nanoporous materials have been utilized in varous research fileds. Especially, nanoporous materials can be used in numerous environmental applications including water treatment, air purification, adsorption of toxic materials etc. So far, many strategies have been developed in order to enhance the properties and performances of functional nanoporous nanomaterials in water treatment clean technology. In this presentation, I will present the interesting results for water treatment clean technology based on design and synthesis of nanoarchitectured nanoporous materials of carbons and metal oxides, and hybrid materials focusing on precursor control, high yield synthesis, activation, shape/orientation control, hybridization with other functional materials. The promising applications in water treatemnt using the developed high-performance porous materials will be presented.

Invited Talk: Sustainable Use of Resources: Production of Food Waste-Based Feeds for Fish Culture

ENGINEERING SUSTAINABLE DEVELOPMENT

Conference: Technical and Engineering Challenges of Addressing Sustainable Development

Year: 2019

Proceeding: Engineering Sustainable Development 2019

Group: Abstract Submissions

Session: Invited Parallel Session 1B - Water Sustainability and Security

Time: Thursday, December 12, 2019 - 12:10pm-12:35pm

Authors:

Man, Y. B., The Education University of Hong Kong
Wong, M. H., The Education University of Hong Kong
Zhang, F., The Education University of Hong Kong
Mo, W. Y., The Education University of Hong Kong

There is about 3,648 tonnes per day of food waste generated in Hong Kong and most of them still end up in our overloaded landfill sites. This is not sustainable as dumping food waste into landfill sites will definitely shorten the lifespan of existing landfill sites. Recycling of food waste as a source of protein to replace part of the fishmeal in commercial fish feed for culturing fish is an alternative way to reuse food waste which could partially ease the disposal pressure. In our studies, a polyculture culture set up was used to grow low trophic level fish, including a filter feeder (bighead, *Aristichthys nobilis*), a herbivore (grass carp,*Ctenopharyngodon idellus*) and a bottom feeder (mud carp, *Cirrhina molitorella*), with lower nutritional requirements. The results of the half-year field trial showed that grass carp and bighead carp fed by food waste-based diets were relatively free of mercury and polycyclic aromatic hydrocarbons (PAHs), compared with those fed with the commercial diets, containing fish meal. Furthermore, food waste-based diets were used for culturing jade perch (*Scortum barcoo*) and Nile tilapia (*Oreochromis niloticus*) in a fish pond for half a year. Health risk assessments based on the concentrations and of metal/loids (arsenic, cadmium, chromium, copper, mercury, lead and zinc) and PAHs in fish meat showed that the culture fish were safe for human consumption. This demonstrated that food waste possesses a high potential as protein source to substitute part of fish meal commonly used in ï¬sh feed for sustainable fish farming.

Invited Talk: Synthesis of Recyclable DNA Triblock Copolymers

Conference: Technical and Engineering Challenges of Addressing Sustainable Development

Year: 2019

Proceeding: Engineering Sustainable Development 2019

Group: Abstract Submissions

Session: Invited Parallel Session 1B - Water Sustainability and Security

Time: Thursday, December 12, 2019 - 12:35pm-1:00pm

Authors:

Li, S., Korea Advanced Institute of Science and Technology
Shin, J., Korea Advanced Institute of Science and Technology

Hybrid polymers containing both biological and synthetic segments are advanced materials that have attracted much interest in recent years. In particular, DNA containing block copolymers can find utility in a wide range of biomedical applications. However, synthesis of these hybrid materials, especially ones with complex chain structures remains to be a major challenge. In this contribution, we report the use of a combination of chemical and biological tools to synthesize DNA containing triblock copolymers. The triblocks have all blocks connected by covalent bonds, thus show superior stability against environmental factors that can denature DNA. Furthermore, we incorporate multiple cloning site (MCS) into the DNA block, and show that the different restriction sites can be effectively cut to regenerate diblock copolymers. The recycled diblocks are then reconnected to create new DNA triblocks of different compositions. This recycling process in principle can be repeated many times thus allowing a wide range of block copolymer structures to be prepared from a limited stock of precursor polymers. Collectively, our study presents a new method to prepare DNA triblock copolymers as well as an effective strategy to recycle the block components in generating new materials.

Invited Talk: Sustainability of Water Management Practices in Hydraulic Fracturing

Conference: Technical and Engineering Challenges of Addressing Sustainable Development

Year: 2019

Proceeding: Engineering Sustainable Development 2019

Group: Abstract Submissions

Session: Invited Parallel Session 1B - Water Sustainability and Security

Time: Thursday, December 12, 2019 - 1:00pm-1:25pm

Authors:

Alessi, D., University of Alberta
Zolfaghari, A., University of Alberta

Hydrocarbons production from combined vertical drilling and hydraulic fracturing (HF) has been touted as a step towards reduced carbon emissions and sustainable and renewable energy technologies. Indeed, greenhouse gas emissions in the United States have declined by 12.2% since 2011, with the United Nations Intergovernmental Panel on Climate Change (IPCC) noting in its 2018 report that most of the decrease is attributable to increased natural gas production from HF. Among the most greenhouse gas intensive parts of HF development is its water cycle, including transport of water to the wellpad to make injected hydraulic fracturing fluids, and the transportation and disposal of saline flowback and produced water (FPW) that is generated for years after production begins. Great strides have been made in the past decade to increase the efficiency of water use, including centralized water hubs and increased treatment, recycling and reuse of FPW for subsequent fracturing jobs. In this presentation, we will discuss the sustainability of the HF water cycle across all major unconventional basins in North America, and the promise for further water use reductions in these regions. Most extant literature views water sustainability from either total volumes of water used or produced, or the salinity of the produced water and ramifications for treatment and reuse costs. Using the Montney play (Alberta, Canada) as a model example, we will further introduce the concept of total produced salts (TPS), as an improved proxy for water sustainability at the play scale in North America and beyond.

Invited Talk: Adsorption Technology for Sustainable Energy and Environmental Applications

ENGINEERING
SUSTAINABLE
DEVELOPMENT

Conference: Technical and Engineering Challenges of Addressing Sustainable Development

Year: 2019

Proceeding: Engineering Sustainable Development 2019

Group: Abstract Submissions

Session: Invited Parallel Session 2A - Sustainable Energy

Time: Thursday, December 12, 2019 - 2:00pm-2:25pm

Authors:

Shang, J., City University of Hong Kong

Chemical separation plays a vital role in the societal development, accounting for 15% annual total energy consumption globally. Typical examples are natural gas processing, oil refinery, and environmental remediation, all of which play important roles in building a sustainable society. An efficient alternative technology to conventional distillation would bring huge benefits though reducing energy use, emissions, and pollution. The gas industry is a 500-billion-dollar input to nearly every sector of the global economy, with high-density storage for transportation and delivery being the most technologically challenging bottleneck. Typical gas storage applications include fuel gases (e.g., methane and hydrogen) for automobiles, natural gas for long-distance transportation, therapeutic medical gases (e.g., oxygen and nitric oxide) for clinical applications, instrument gases (e.g., nitrogen and argon) for industry usage, and electronic gases (e.g., arsine and phosphine) delivery in the semiconductor fabrication processes. Adsorption technology using porous materials can offer highly efficient routes for gas separation and storage applications. In this talk, I will introduce the basics of adsorption technology and present our latest accomplishment, especially the concept of active sieving technology, which enables highly efficient and economical molecular separation and storage not possible before.

Invited Talk: Exploitation of Advanced Materials and Process for Commercialization of Perovskite Solar Cell

ENGINEERING
SUSTAINABLE
DEVELOPMENT

Conference: Technical and Engineering Challenges of Addressing Sustainable Development

Year: 2019

Proceeding: Engineering Sustainable Development 2019

Group: Abstract Submissions

Session: Invited Parallel Session 2A - Sustainable Energy

Time: Thursday, December 12, 2019 - 2:25pm-2:50pm

Authors:

Jung, H. S., Sungkyunkwan University (SKKU)

All solid-state solar cells based on organometal trihalide perovskite absorbers have already achieved distinguished power conversion efficiency (PCE) to over 25% and further improvements are expected up to 27%. Now, the research on perovskite solar cells has been moving toward commercialization. To facilitate commercialization of these great solar cells, some technical issues such as long-term stability, large scale fabrication process, and Pb-related hazardous materials need to be solved. Also, flexible perovskite solar cell using plastic substrate can be used in niche applications such as portable electrical chargers, electronic textiles, and large-scale industrial roofing.

This talk is dealing with our recent efforts to facilitate commercialization of perovskite solar cells. For examples, we introduce a recycling technology of perovskite solar cells, which covers the regeneration process of Pb contained perovskite layer as well as recycling process of Au electrodes and transparent conducting oxide glass. Also, recent interesting results regarding ultra-flexible perovskite cells will be discussed in terms of stress analysis. Finally, simple fabrication technologies for realizing large scale perovskite module is going to be introduced.

Invited Talk: Rational Design of Manganese Based Layered Oxides for Sodium Ion Batteries

ENGINEERING
SUSTAINABLE
DEVELOPMENT

Conference: Technical and Engineering Challenges of Addressing Sustainable Development

Year: 2019

Proceeding: Engineering Sustainable Development 2019

Group: Abstract Submissions

Session: Invited Parallel Session 2A - Sustainable Energy

Time: Thursday, December 12, 2019 - 3:15pm-3:40pm

Authors:

Kang, Y. M., Dongguk University-Seoul

Phase transformation of layered structure into spinel structure has been detrimental for most of layered oxide cathodes.1 Even if a lot of efforts have been made to relieve this highly irreversible phase transformation, there have been few successful results. However, we firstly observed the possibility to make this irreversible phase transformation extremely reversible by utilizing Na-birnessite (Na_xMnO_2 â€¢yH_2O; Na-bir) as a basic structural unit, which has distinctive layered structure containing crystal water. Herein, the crystal water in the structure contributes to generating metastable spinel-like phase, which is the key factor for making this unusual reversibility happen.2 The reversible structural rearrangement between layered and spinel-like phases during electrochemical reaction could activate new cation sites and enhance ion diffusion with higher structural stability. This unprecedented reversible phase transformation between spinel and layered structure was deeply analyzed via combined ex situ soft and hard X-ray absorption spectroscopy (XAS) analysis with in situ X-ray diffraction (XRD). Fundamental mechanism on this reversible phase transformation was theoretically elucidated and confirmed by kinetic investigation using first-principle calculation. These results provide deep insight into novel class of intercalating materials which can deal with highly reversible framework changes, and thus it can pave an innovative way for the development of cathode materials for next-generation rechargeable batteries.

Invited Talk: Low Impact Development for Stormwater Treatment in Densely Populated Areas

ENGINEERING
SUSTAINABLE
DEVELOPMENT

Conference: Technical and Engineering Challenges of Addressing Sustainable Development

Year: 2019

Proceeding: Engineering Sustainable Development 2019

Group: Abstract Submissions

Session: Invited Parallel Session 2B - Economic, Social, Climatic and Environmental Issues

Time: Thursday, December 12, 2019 - 2:00pm-2:25pm

Authors:

Tsang, D. C., The Hong Kong Polytechnic University

Bioretention systems, one of the best management practices for low impact development for water recycling/reuse, are required for a substantial removal of various heavy metals for non-potable reuse. This study investigated the efficiency of food waste compost and waste wood-derived biochar for heavy metal removal from synthetic stormwater under intermittent flow. The heavy metal removals by compost and biochar were promising throughout the 84 pore volumes of stormwater infiltration. Only a small portion of heavy metals attached on the compost was remobilized during the drainage phase of intermittent flow. In comparison with geomedia sorbent, compost and biochar are effective in retarding heavy metal breakthrough. Although the biochar showed a specific surface area approximately 1.5 times of compost, a substantial increase in surface oxygen-containing functional groups was found with compost, which contributed to the higher removal capacity and persistence against drainage. The co-existing kaolinite enhanced heavy metal removal due to plentiful active sites, whereas humic acid facilitated their transport *via* humate-metal complexation. Nevertheless, in view of the more recalcitrant environmental stability and customable physiochemical properties, biochar appeared to give better performance in pilot- or field-scale bioretention systems for stormwater harvesting.

Invited Talk: Pollution Patterns, Characteristics and Risks of Perfluorinated Compounds in Surface Water in Taiwan

ENGINEERING
SUSTAINABLE
DEVELOPMENT

Conference: Technical and Engineering Challenges of Addressing Sustainable Development

Year: 2019

Proceeding: Engineering Sustainable Development 2019

Group: Abstract Submissions

Session: Invited Parallel Session 2B - Economic, Social, Climatic and Environmental Issues

Time: Thursday, December 12, 2019 - 2:25pm-2:50pm

Authors:

Jiang, J. J., Chung Yuan Christian University
Okvitasari, A., Chung Yuan Christian University

Perfluorinated compounds (PFCs) are synthetic and environmentally persistent pollutants that are amenable to transport and accumulation in aquatic environment. In this study, the water samples were collected from June 2017 to May 2018, and among 9 PFCs were investigated in the Nankan River in northern Taiwan. PFOA and PFOS were the two most dominant compounds with a mean concentration of 24.02 ng/L and 9.45 ng/L, respectively. Principle component analysis (PCA) and positive matrix factorization (PMF) model were employed to quantitatively calculate the contributions of sources. A total 40% (Factor 1) of the PFCs in the Nankan River was estimated to originate from various municipal activities source, including used in aqueous film-forming (AFF), 36% (Factor 2) from chemical industry and semiconductor and photonic industrial wastewater, and 24% (Factor 3) from industrialized and residential areas, such as household cleaning products, pesticides and insecticides, food packaging and originate from pesticide and chemical factories. The health risk of PFCs from consuming crops irrigated by Nankan River was assessed and the health risk assessment model was less than 1, showing that no obvious adverse effects to human health.

Invited Talk: A Case Study on the Use of Existing Buildings As a Sustainable, Cost-Effective Solution for Affordable Housing

ENGINEERING
SUSTAINABLE
DEVELOPMENT

Conference: Technical and Engineering Challenges of Addressing Sustainable Development

Year: 2019

Proceeding: Engineering Sustainable Development 2019

Group: Abstract Submissions

Session: Invited Parallel Session 2B - Economic, Social, Climatic and Environmental Issues

Time: Thursday, December 12, 2019 - 2:50pm-3:15pm

Authors:

Delia, R. C., Housing Joint Venture

As populations continue to migrate into cities, housing affordability and long term infrastructure sustainability become pressing issues. In USA 31.5% of households are considered cost burdened, meaning they pay more than 30% of their income towards housing. In urban communities, that number rises to 46 percent.

Interestingly, some urban neighborhoods also possess a clustering of vacant, distressed properties. The collocation of economic issues with viable cost-effective infrastructure is an opportunity to meet the social challenge.

By extending the useful life of existing structures through renovation, economic and social resiliency can be introduced to otherwise overlooked neighborhoods. The embodied energy from the initial construction is put back into service and the building is rescued from functional obsolescence with conscious renovation. The purchase, renovate to rent model allows a targeted approach by focusing on gaining a critical mass in a specific geography. Furthermore, a greater production of housing units at all price points could play a pivotal role in reversing the longer-term rise in housing cost burdens.

In this work, we offer a case study of individual properties that function as catalyst for change in so called â€œGrowth Value Neighborhoodsâ€. Moderate, functional renovations that leverage existing materials prevent additional waste to landfill while simultaneously addressing the housing shortage plaguing the midwest United States.

Invited Talk: Sustainable Upcycling of Low-Value Precursors to Porous Carbon Materials for CO2 Capture

ENGINEERING SUSTAINABLE DEVELOPMENT

Conference: Technical and Engineering Challenges of Addressing Sustainable Development

Year: 2019

Proceeding: Engineering Sustainable Development 2019

Group: Abstract Submissions

Session: Invited Parallel Session 2B - Economic, Social, Climatic and Environmental Issues

Time: Thursday, December 12, 2019 - 3:15pm-3:40pm

Authors:

Lee, K. B., Korea University
Yuan, X., Korea University
Choi, S. W., Korea University

The increase of greenhouse gas emissions is causing global warming and the related climate change issues worldwide. CO_2 capture and sequestration (CCS) is considered to be a promising strategy for reducing CO_2 emissions. Adsorption using solid materials is promising for CO_2 capture because this approach exhibits advantages of low energy requirement for regeneration, scale-up feasibility, and mild operating conditions. Among various adsorbents, activated carbon is attractive because of advantages such as cost-effectiveness, chemical and mechanical stability, and tunable pore structure. Activated carbon can be obtained from carbon-containing precursors. In this study, low-value carbon precursors were used to produce high value-added porous carbon for capturing greenhouse gases including CO_2. From the experimental results, the relation between porosities and CO_2 adsorption uptakes were investigated and the pore size that is crucial for CO_2 capture was determined. The prepared porous carbons not only exhibited a high CO_2 uptake, but also good selectivity, simple regeneration, excellent cyclic stability, and rapid adsorption-desorption kinetics.

Techno-Economic Feasibility of Distributed Waste-to-Hydrogen Systems to Support Green Transport in Glasgow

Conference: Technical and Engineering Challenges of Addressing Sustainable Development

Year: 2019

Proceeding: Engineering Sustainable Development 2019

Group: Abstract Submissions

Session: Parallel Session 3A - Sustainable Energy

Time: Thursday, December 12, 2019 - 3:35pm-4:00pm

Authors:

You, S., University of Glasgow
Lui, J., University of Glasgow

Significant greenhouse gas and pollutant emissions from fossil-fuelled transport call for alternative green transport fuels like hydrogen. Biomass waste is a low carbon source for hydrogen production via various thermochemical and biochemical technologies. As a result, distributed Waste to Hydrogen (WtH) systems are a potential solution to tackle the twin challenges of sustainable waste management and transport. In this work, a novel concept of distributed WtH systems based on gasification/fermentation to support hydrogen fuel cell buses in Glasgow was proposed. The system configurations of WtH systems were identified for the different technology routes, based on hydrogen production process modelling that includes the choice of biomass waste feedstock, hydrogen production reactors, and upstream and downstream system components. In the process modelling, the impacts of operating conditions on the hydrogen production were evaluated to develop the optimal process parameters based on maximum hydrogen yields. A cost-benefit analysis was conducted to evaluate the economic feasibility of the systems, based on the accumulation of a systematic database including direct cost data on construction, maintenance, operations, materials, transport and storage, along with indirect cost data comprising environmental impacts and externalities, cost of pollution, carbon taxes and subsidies. Sensitivity analysis was conducted to identify the most significant factors determining the profitability of distributed WtH systems supporting hydrogen fuel cell buses in Glasgow.

Electrochemical Coupling of Aromatic Alcohols and Carbon Dioxide for Electrosynthesis of ?-Hydroxy Acids in a Sustainable Way

Conference: Technical and Engineering Challenges of Addressing Sustainable Development

Year: 2019

Proceeding: Engineering Sustainable Development 2019

Group: Abstract Submissions

Session: Parallel Session 3A - Sustainable Energy

Time: Thursday, December 12, 2019 - 4:00pm-4:25pm

Authors:

Kim, M., Korea Institute of Energy Research
Muchez, L., KU Leuven
Vos, D. D., KU Leuven

The chemical incorporation of CO_2 into organic molecules will be an effective approach for sustainable low carbon society. Carboxylation is one of the best way to make C-C bond through the reaction of CO_2 with organic molecules. However, CO_2 is highly thermodynamically stable, in order to activate it, carboxylation often demands harsh reaction conditions, such as high temperature and high pressure, and environmentally harmful organometallic catalysts. An electrochemical route for carboxylation can be free from these issues by using electrons as a redox reagent. The electrosynthesis of carboxylic acid is a highly promising method in organic synthesis from the viewpoint of green chemistry and CO_2 utilization.

For a sustainable electrosynthesis of carboxylic acid, herein, we report a novel methodology operating in a simple undivided cell with a non-sacrificial anode, at which tetramethylpiperidine-1-oxyl (TEMPO)-mediated alcohol oxidation takes place. Aromatic alcohols can be oxidized with high efficiency to ketones or aldehydes at the anode in the presence of a small amount of water, and then converted to Î±-hydroxy acids in yields of up to 61 % by coupling with CO_2 at the cathode. The electrocarboxylation mechanism comprising electron and proton transfers at the cathode is discussed along with experiments using various cathode materials. Electron and proton transfers are influenced by the hydrogen adsorption energy of the cathode material, which plays an important role in the electrocarboxylation efficiency.

Bioalcohol Production from Acidogenic Products

Conference: Technical and Engineering Challenges of Addressing Sustainable Development

Year: 2019

Proceeding: Engineering Sustainable Development 2019

Group: Abstract Submissions

Session: Parallel Session 3A - Sustainable Energy

Time: Thursday, December 12, 2019 - 4:25pm-4:50pm

Authors:

Jung, S., City College of New York
Kwon, E., Sejong University

In an effort to seek a sustainable platform for bioalcohol production, the technical hybridization of the fermentation and thermo-chemical processes have been made. Specifically, acidogenic products (*i.e.*,volatile fatty acids: VFAs) from the fermentation process of food waste were used as an initial feedstock, and the thermo-chemical conversion of VFAs into bioalcohols was done. In particular, this study experimentally proved that the yield of bioalcohol derived from VFAs reached up to 45 % by applying the biochar composite as a catalyst. Note that biochar was fabricated from co-pyrolysis of chitin and oyster shell. One of the innovative technical features in line with bioalcohol conversion in this study was that the conversion of bioalcohol was achieved by a single conversion process, and its conversion did not take more than 1 second. The reported conversion mechanism for bioalcohol has not been yet optimized. Nevertheless, all experimental findings in line with bioalcohol conversion in this study could be a viable route for valorizing food waste dye to the technical simplicity.

GVL Pretreatment for Enzymatic Hydrolysis and Lignin Extraction from Lignocellulosic Biomass

ENGINEERING
SUSTAINABLE
DEVELOPMENT

Conference: Technical and Engineering Challenges of Addressing Sustainable Development

Year: 2019

Proceeding: Engineering Sustainable Development 2019

Group: Abstract Submissions

Session: Parallel Session 3A - Sustainable Energy

Time: Thursday, December 12, 2019 - 4:50pm-5:15pm

Authors:

Ahmed, M. A., Seoul National University pyeongchang campus
Choi, J. W., Seoul National University

In this study we adapted ball milled biomass as a model substrate in the bench type batch reactors for the pretreatment to observe effects on enzymatic hydrolysis and lignin yield. The experiments were intitally conducted in an oil bath for the conditions of temperature of (60-100 °C), process time of (24- 48 Hrs) & the Aquous GVL concentration was fixed to 80% for all the experiments. Initially, the pinewood was ball milled was different time intervals and 20 Hrs of milling time was screened out to be carried for all the experiments. The lignin yield increased from 3% to 8% when the experiments were performed from 60 C : 24 Hrs to 100 C: 48 Hrs. The analysis of the pretreated biomass lspecifically interms of its composition was carried out. Also the enzymatic digestibility of the fractionated biomass showed a noticeable correlation with the final glucose release . From here it can be conclusively suggested that GVL can be one potential option for the pretreatment of lignocellulosic biorefineries.

Visualization of Nickel Sulfide Pulverizing in Hierarchical Carbons; Applied for the Anode of Lithium Ion Battery

ENGINEERING SUSTAINABLE DEVELOPMENT

Conference: Technical and Engineering Challenges of Addressing Sustainable Development

Year: 2019

Proceeding: Engineering Sustainable Development 2019

Group: Abstract Submissions

Session: Parallel Session 3A - Sustainable Energy

Time: Thursday, December 12, 2019 - 5:15pm-5:40pm

Authors:

Park, J. H., Korea Advanced Institute of Science & Technology (KAIST)
Lee, J. W., Korea Advanced Institute of Science & Technology (KAIST)

Metal sulfides have been spotlighted as next-generation anode materials of lithium ion battery due to their high theoretical capacity. In addition, they have better thermal stability and electrical conductivity than metal oxide series while having similar theoretical capacity. Among them, the nickel sulfide (NiS) has high theoretical capacity of 590 mAh g^{-1}. However, when NiS was used for the anode material of lithium ion battery, it was pulverized in a long term operation, and electrochemically isolated eventually. To solve this problem, this work suggested the synthesis of the NiS coated by hierarchical carbon shell. The NiS was mixed with $CaCO_3$ nano-templates and then coated with carbon using a toluene bubbling CVD. As a result, hierarchical structured carbon was synthesized on the surface of inner NiS. The resultant material shows a high capacity of 575 mAh g^{-1} at 0.1 A g^{-1} and maintain over 483 mAh g^{-1} at 200th cycle for the anode material of lithium ion battery. During this long-term cycling, inner NiS was pulverized in the hierarchical carbon, which could contain pulverized pieces until the 100th cycle. However, after 100 cycles, the pulverized NiS was separated into Ni and Li_2S, which leads to the formation of polysulfide. This converted polysulfide has high solubility and mobility in the organic electrolyte, and eventually diffused out of the carbon coating. This process was confirmed by the cycle-by-cycle TEM images and observed using additional ex situ SEM and XRD analyses.

Atmospheric Water Extractor and Space Cooling Unit

ENGINEERING SUSTAINABLE DEVELOPMENT

Conference: Technical and Engineering Challenges of Addressing Sustainable Development

Year: 2019

Proceeding: Engineering Sustainable Development 2019

Group: Abstract Submissions

Session: Parallel Session 3A - Sustainable Energy

Time: Thursday, December 12, 2019 - 5:40pm-6:05pm

Authors:

Khan, M. S., National University of Singapore
Ali, W., Jizan University

Water is the elixir of life and without it no life can sustain on planet earth. There are many people in the world without access to clean water and are facing ailments of water-borne diseases. A lot of efforts are going on to eradicate potable water shortage but it keeps getting bigger with rapid urbanization and increasing population. The problem is particularly severe in hot and humid climates that have the perfect receipt for extracting water from air using solar energy. In this work the design of potable solar-powered water extracting and space cooling device is discussed. The designed device has the potential to keep running for 20 years through a square meter of photovoltaic solar panel extracting potable water at an average rate of 35 ml/hr @ 60% RH while maintaining the temperature of 2 cu.ft. space between 12-15°C. The design employed Peltier modules for bringing ambient air to it its dew point. The design was experimentally investigated at different relative humidity and the amount of water generated found to be strongly related to RH though small dependence also is seen with inlet airflow/temperature. The design has the potential to provide water in arid and off-grid conditions particularly reliable during natural disasters. The space cooling part can be used to store important medicines during power grid failure. The use of device reduce strain on conventional potable water supply and create awareness and utilization of abundant solar energy while keeping zero emissions.

Flexible Polyvinyl Alcohol/Reduced Graphene Oxide Coated Carbon Composites for Electromagnetic Interference Shielding

ENGINEERING
SUSTAINABLE
DEVELOPMENT

Conference: Technical and Engineering Challenges of Addressing Sustainable Development

Year: 2019

Proceeding: Engineering Sustainable Development 2019

Group: Abstract Submissions

Session: Parallel Session 3A - Sustainable Energy

Time: Thursday, December 12, 2019 - 6:05pm-6:30pm

Authors:

Lai, D., Institute of Urban Environment, Chinese Academy of Sciences

The proliferating market of electronic devices has driven the demand for efficient electromagnetic interference (EMI) materials to eliminate unwanted electromagnetic radiation. Despite great recent progress with graphene-based polymer composites, the facile preparation of cost-efficient and high-performance EMI shielding composites at low graphene loading has yet to be achieved. Herein, flexible polyvinyl alcohol/reduced graphene oxide coated activated carbon (PVA/RGO@AC) composite films with extremely low graphene amounts are prepared by using AC as segregators and substrates. Decoration of AC with graphene to create an individual RGO sheet coated AC structure leads to a dramatic increase in the conductivity of AC and effectively prevents the restacking and agglomeration of graphene. The percolation threshold of the PVA/RGO@AC composites is as low as 0.17 wt% for RGO@AC, and in particular, only 0.017 wt% RGO is needed. A high conductivity of 10.90 S/m and impressive electromagnetic interference shielding effectiveness (EMI SE) of 25.6 dB with an absorption-dominated mechanism are achieved for PVA/RGO@AC composites with a low RGO loading of 1.0 wt%. The specific EMI SE of the composite reaches 17.5 dB/mm, outperforming most of the reported pioneering graphene-based polymer composites with such low RGO amount. The excellent electrical property and outstanding EMI shielding performance are attributed to the internal well-constructed three-dimensional RGO-AC-RGO interconnected conductive network. Intriguingly, the fabricated composites exhibit a stable EMI SE even after 1000 bend-release cycles. These results demonstrate that our approach is a novel and promising method for producing highly conductive, high shielding performance and cost-effective materials with very low graphene loading.

Water Compatible Conjugated Polymer Photocatalysts By Side Chain Engineering

Conference: Technical and Engineering Challenges of Addressing Sustainable Development

Year: 2019

Proceeding: Engineering Sustainable Development 2019

Group: Abstract Submissions

Session: Parallel Session 3B - Water Sustainability and Security

Time: Thursday, December 12, 2019 - 3:35pm-4:00pm

Authors:

Byun, J., Korea Institute of Science and Technology

Conjugated polymers have emerged as a new class of heterogeneous photocatalysts for utilizing visible light as an environmentally friendly energy source. By virtue of structural designability, the conjugated polymers can be tuned to show better hydrophilicity by side chain engineering particularly for targeting the photocatalytic reactions in an aqueous medium, i.e. photocatalytic degradation of contaminants. We have designed a linear chain conjugated polymer to have different terminal functionalities such as diethylamine (DEA) and vinylimidazole (Vim) for initiating a reversible wettability change in water by the introduction of external stimuli. The DEA-based conjugated polymer exhibited the reversible hydrophilicity in the presence of CO_2 gas as the tertiary amine captured CO_2 molecules with water to form ammonium-bicarbonate salt, making the whole polymers hydrophilic for improving photocatalytic activity in water. Once the CO_2 was desorbed, the polymer could be easily separated from water for recycling. In the case of Vim-terminated conjugated polymer, the polymer chain could form a polyelectrolyte complex with hydrophilic polymer by ionic complexation, giving a hydrogel photocatalyst to show great volume expansion in water. With the structural expansion, the photoactive sites in hydrogel were also expanded, enhancing the overall photocatalytic activities in aqueous solution. The utility of the water-compatible conjugated polymer photocatalysts was demonstrated under visible light illumination, namely, photodegradation of organic dyes and micropollutants in water.

Adsorption of Methylene Blue, Copper, and Cadmium Onto Activated Carbon Prepared from Hydrothermal Carbonization of Teak (Tectona grandis) Sawdust

ENGINEERING
SUSTAINABLE
DEVELOPMENT

Conference: Technical and Engineering Challenges of Addressing Sustainable Development

Year: 2019

Proceeding: Engineering Sustainable Development 2019

Group: Abstract Submissions

Session: Parallel Session 3B - Water Sustainability and Security

Time: Thursday, December 12, 2019 - 4:00pm-4:25pm

Authors:

Nguyen, H. D., Institute of Environmental Engineering, National Central University
Tran, H. N., Institute of Fundamental and Applied Sciences, Duy Tan University
Lin, C. C., Institute of Environmental Engineering, National Central University
Chao, H. P., Department of Environmental Engineering, Chung Yuan Christian University

Due to the concurrent industrialization and population growth in recent years, Vietnam and other developing countries have been facing increasing problems of environmental pollution, such as water contamination with dyes and heavy metals. Among the methods for removing contaminants, adsorption is regarded as a relatively low-cost and effective means. In this regard, development of carbonaceous adsorbents via a more straightforward, greener, and robust way is essential. Here, activated carbon (AC) samples synthesized from teak sawdust through hydrothermal carbonization and chemical activation were investigated for the adsorption of methylene blue (MB), Cd(II), and Cu(II). The sawdust was carbonized by a hydrothermal process and then activated using the activating agents K_2CO_3 or $ZnCl_2$ in various weight ratios. The carbonaceous adsorbents were characterized by SEM, S_{BET}analyzer, FTIR, XPS, and Boehm titration method. Characterization results show that AC samples possessed high S_{BET} and rich oxygen-containing functional groups. The maximum adsorption capacities obtained for MB, Cd(II), and Cu(II) were 516 mg/g, 166 mg/g, and 159 mg/g, respectively. The amount of oxygen-containing functional groups is regarded as an important factor in determining the adsorptive amounts. Electrostatic force was found to be the primary adsorption mechanism for the contaminants tested. The complexation reaction is a vital adsorption mechanism for Cu^{2+}. Other mechanisms are less important in the adsorption of MB and Cd^{2+}. The high adsorption capacity of AC for MB, Cu, and Cd suggests that hydrothermal carbonization followed by chemical activation process has the potential application to produce efficient adsorbents used in the wastewater treatment process.

Groundwater Study in Coastal Region: Hydrogeology and Geochemical Characteristic of an Island in Malacca Strait, Indonesia

ENGINEERING
SUSTAINABLE
DEVELOPMENT

Conference: Technical and Engineering Challenges of Addressing Sustainable Development

Year: 2019

Proceeding: Engineering Sustainable Development 2019

Group: Abstract Submissions

Session: Parallel Session 3B - Water Sustainability and Security

Time: Thursday, December 12, 2019 - 4:25pm-4:50pm

Authors:

Putra, D. B. E., Universitas Islam Riau
Hadian, M. S. D., Universitas Padjadjaran
Yuskar, Y., Universitas Islam Riau
Syah Alam, B. Y. C., Universitas Padjadjaran
Suryadi, A., Universitas Islam Riau
Wan Yaacob, W. Z., Universiti Kebangsaan Malaysia
Datta, B., James Cook University

Groundwater is the substantial natural resource in Indonesia as it is become the major source of clean water used for drinking and irrigation activities. However, both quality and quantity of this important resource has facing depletion caused by several anthropogenic activities such as excessive pumping, industrial activities and land use changing. In the coastal area, additional problem to the groundwater resource generally had been found contributed by saltwater intrusion. Bengkalis Island located in the Malacca strait is one of the important regions that connecting Indonesia and Malaysia. The area is developing particularly along the coast, those condition eventually resulting in increased of human activities and affecting the amount of clean water demand. Most people in the study area use dug well to obtain groundwater for their daily need, but brackish water had been discovered in the wells that located near the shore as well as in the distance. Geological information of groundwater aquifer was determined from 4 deep boreholes (±80m each) and 12 lines of electrical geophysics. Groundwater level measured from dug as well as bore wells to mapped groundwater flow, several groundwater samples analyzed in laboratory to specify its chemical content used in determination of groundwater quality and isotope examination for groundwater dating. Furthermore, both isotope and electrical geophysics data was used to trace saltwater intrusion phenomenon and boundary. The result of this research project will be use as scientific evidence by local government in regional development, establishing regulation for groundwater and coastal area management to achieve groundwater sustainability.

Inadequacy of Using Individual Pollutants to Assess Comprehensive Hazard Effects of Road Deposited Sediments: Implications for Stormwater Reuse Safety

ENGINEERING SUSTAINABLE DEVELOPMENT

Conference: Technical and Engineering Challenges of Addressing Sustainable Development

Year: 2019

Proceeding: Engineering Sustainable Development 2019

Group: Abstract Submissions

Session: Parallel Session 3B - Water Sustainability and Security

Time: Thursday, December 12, 2019 - 4:50pm-5:15pm

Authors:

Liu, A., Shenzhen University
Hong, N., Shenzhen University
Yang, M., Shenzhen University

Road deposited sediments (RDS) are the key carrier of pollutants deposited on road surfaces during dry days. When rainfall events occur, RDS can be washed-off into stormwater runoff along with pollutants attached, undermining stormwater reuse safety. The conventional approach to assessing the hazard effect of pollutants attached to RDS in terms of stormwater reuse is to analyze individual pollutant groups and their quantity. However, it is well known that many pollutants are present together rather than individually. This raises a question: do conventional approaches permit a comprehensive understanding of how appropriately the RDS polluted stormwater can be reused?

In order to answer this question, this study undertook a toxicity test of pollutants attached to RDS using Chinese hamster ovary cells (CHO). Organic pollutants were used as the testing pollutant group. It is noted that comparative toxicity of RDS is not strongly related to total solids (commonly seen as the key carrier of pollutants) and chemical oxygen demand (COD, representing organic matters). Additionally, the comparison results of spatial distributions of toxicity (in this study) and individual pollutants in previous studies did not show a similar trend. These results imply that toxicity should be also used to indicate how stormwater can be safely reused while solely investigating individual pollutants can not adequately show a comprehensive hazard effect in terms of ensuring stormwater reuse safety. Based on study outcomes, a new assessment approach considering both pollutant and toxicity were proposed. This will assist on effective stormwater reuse and ensuring their reuse safety.

Magnetic Biosorbents: Their Preparation, Application, and Regeneration for Wastewater Treatment

ENGINEERING
SUSTAINABLE
DEVELOPMENT

Conference: Technical and Engineering Challenges of Addressing Sustainable Development

Year: 2019

Proceeding: Engineering Sustainable Development 2019

Group: Abstract Submissions

Session: Parallel Session 3B - Water Sustainability and Security

Time: Thursday, December 12, 2019 - 5:40pm-6:05pm

Authors:

Hassan, M., University of Newcastle

The utilisation of magnetic biosorbents (metal or metal nanoparticles impregnated onto biosorbents) has attracted increasing research attention due to their manipulable active sites, specific surface area, pore volume, pore size distribution, easy separation, and reusability that are suitable for remediation of heavy metal(loid)s and organic contaminants. The properties of magnetic biosorbents (MB) depend on the raw biomass, properties of metal nanoparticles, modification/synthesis methods, and process parameters which influence the performance of removal efficiency of organic and inorganic contaminants. Ion exchanges, electrostatic interaction, precipitation, and complexation are the dominant sorption mechanisms for ionic contaminants whereas hydrophobic interaction, interparticle diffusion, partition, and hydrogen bonding are the dominant adsorption mechanisms for removal of organic contaminants by magnetic biosorbents. In generally, low pyrolysis temperatures are suitable for ionic contaminants separation, whereas high pyrolysis temperatures are suitable for organic contaminants removal. Additionally, magnetic properties of the biosorbents are positively correlated with the pyrolysis temperatures. Metal-based functional groups of MB can contribute to an ion exchange reaction which influences the adsorption capacity of ionic contaminants and catalytic degradation of non-persistent organic contaminants. Metal modified biosorbents can enhance adsorption capacity of anionic contaminants significantly as metal nanoparticles are not occupying positively charged active sites of the biosorbents. Magnetic biosorbents are promising adsorbents in comparison with other adsorbents including commercially available activated carbon, and thermally and chemically modified biochar in terms of their removal capacity, rapid and easy magnetic separation which allow multiple reuse to minimize remediation cost of organic and inorganic contaminants from wastewater.

Performance of Pb(II) Removal By an Activated Carbon Supported Nanoscale Zero-Valent Iron Composite at Ultralow Iron Content

ENGINEERING
SUSTAINABLE
DEVELOPMENT

Conference: Technical and Engineering Challenges of Addressing Sustainable Development

Year: 2019

Proceeding: Engineering Sustainable Development 2019

Group: Abstract Submissions

Session: Parallel Session 3B - Water Sustainability and Security

Time: Thursday, December 12, 2019 - 5:15pm-5:40pm

Authors:

Liu, X., Institute of Urban Environment

Activated carbon (AC) is commonly used as a kind of versatile material for drinking water purification. However, its application is limited by low efficiency and capacity of heavy metals. In addition, it is also concerned that secondary release of adsorbed heavy metals. Therefore, modification of AC is indispensable. In present work, AC supported nanoscale zero-valent iron composite (NZVI/AC) at ultralow iron content was synthesized and used to remove Pb(II) from aqueous solution. The technical characterization demonstrated that the loaded amorphous NZVI nanoparticles had a chain-like shape in or close to pores and were found as individual nanospheres with size of approximately 10 nm on the outer surface. The adsorption capacity of Pb(II) by NZVI/AC was 59.35 mg g^{-1} at 298.15K with a pH of 6.00, which was 8.2 times than that of AC support only. The NZVI/AC with the iron content of only 1.57% showed a highly efficient Pb(II) removal performance with 95% of Pb(II) eliminated within 5 min. The monitoring of iron release indicated no iron was released at a pH above 4.02. The Pb(II) removal by NZVI/AC was well-represented by a pseudo-second-order kinetics model and showed the behavior of an exothermic process. Essentially, Pb(II) was converted to insoluble forms such as Pb^0, $PbCO_3$, $Pb(OH)_2$, PbO or white lead ($[2PbCO_3 \cdot Pb(OH)_2]$). To sum up, these results indicate the first fabrication of NZVI/AC composites with such low iron loading that still present an outstanding Pb(II) removal performance in drinking water purification absence of any potential risk.

Aerobic Granulation of Micrococcus Aloeverae Strain SG002 and Its Application in Oil Remediation and Polyhydroxyalkanoates (PHA) Production

Conference: Technical and Engineering Challenges of Addressing Sustainable Development

Year: 2019

Proceeding: Engineering Sustainable Development 2019

Group: Abstract Submissions

Session: Parallel Session 3B - Water Sustainability and Security

Time: Thursday, December 12, 2019 - 6:05pm-6:30pm

Authors:

Ghosh, S., Indian Institute of Technology Guwahati
Chakraborty, S., Indian Institute of Technology Guwahati

Aerobic granulation technology is an effective outcome in industrial wastewater treatment. Granulation technique became popular for its simple operation with less space and resource consumption having highly settleable compact sludge granules consisting of numerous pollutant tolerant microbial populations. In oily wastewater treatment, mixed sludge aerobic granules often suffered from slow granulation and rapture in high oil concentration. *Micrococcus aloverae* strain SG002 isolated from effluent treatment plant of Indian Oil Corporation Limited, Guwahati, India, featured fast self granulation in an aerobic granular reactor (AGR) while treating complex oily wastewater containing 1000 mg/L chemical oxygen demand (COD) and 250 mg/L emulsified diesel. In 50 days, granule biomass concentration and extracellular polymeric substances reached up to 6.34±0.55 mg/L and 216±2.14 mg/g VSS (volatile suspended solids), respectively. Electron microscopic images revealed dense spherical microbial granule structure. Granules were capable of 99Â±0.23% COD and 95.34Â±1.12% oil removal within 50 days of operation. However, at the end of feast phase granules produced approximately 312Â±10 mg/g VSS biodegradable polymer polyhydroxyalkanoates (PHA) which is considered to be an ecofriendy bioplastic. AGR treated effluent was reused for watering *Cicer arietinum*legume plants. AGR effluent (oil concentration: 12.5Â±0.5 mg/L) proved to be very less phototoxic providing good plant growth with 0.67Â±0.011 mg/g F.W (fresh weight) cholorophyll, 0.11 Âµ mole H_2O_2 split/mg F.W. catalase and 0.056Â±0.002 Î" Absorbance/mg F.W. peroxidase contents after 4 weeks. Hence, this granulation strategy revealed a new scope in petroleum wastewater treatment with potential biomaterial production and water reuse in agricultural irrigation.

Green Engineering Studies on Potential Solvent for the Removal of Endocrine Disrupts Chemicals from Water Matrices: A COSMO-RS Model Based Approach

Conference: Technical and Engineering Challenges of Addressing Sustainable Development

Year: 2019

Proceeding: Engineering Sustainable Development 2019

Group: Abstract Submissions

Session: Parallel Session 3B - Water Sustainability and Security

Time: Thursday, December 12, 2019 - 6:30pm-6:55pm

Authors:

Vivek,, M. S., SSN College of Engineering, Rajiv Gandhi Salai (OMR), Kalavakkam, Tamilnadu-603110.
Anantharaj, R., SSN College of Engineering, Rajiv Gandhi Salai (OMR), Kalavakkam, Tamilnadu-603110.
K, S. K., SSN College of Engineering

A thoughtful study on screening, synthesising and characterising various*Ionic liquids (IL's) and Deep Eutectic Solvents (DES)* for the extraction of endocrine disrupts chemicals (EDCs) like bisphenol A, phenol, Î²-estradiol from water matrices based on COSMO-RS theory. The DFT method will be used to produce more reliable geometries for hydrogen-bond formation in the mixture of solvent and EDCs. The structural properties such as binding energy, orbital energy and partial charge of atoms on the molecules were measured and these are useful descriptors for the strength of hydrogen bond, short-range and long-range interaction between screened solvent and EDCs at molecular and atom level.

Excess thermodynamic properties of binary mixture were determined in order to understand the deviation from ideal solution behaviour and are easily relate activity coefficients. The activity coefficient at infinite dilution describes the non-ideality for chosen species in a mixture based on solute-solvent interactions. Finally, the extraction process was conducted using screened solvent and examined the tie line composition in order to generate the triangular diagram to better understanding of solute-solvent behaviour and determined the efficiency of the separation process at atmospheric conditions. The reliability and consistency of the experimental data will be evaluated by using Othmer-Tobias and Hand equation. The effectiveness of extraction of EDCs from water matrices using ILâ€™s and DES were measured by means of selectivity, distribution coefficient and performance index. On the other hand, the feasibility test was carried out for solvent regeneration and reuse which makes water insensitive solvent attractive process involving multiple cycles.

Microplastics:ALL That Glitters Is NOT Gold

Conference: Technical and Engineering Challenges of Addressing Sustainable Development

Year: 2019

Proceeding: Engineering Sustainable Development 2019

Group: Abstract Submissions

Session: Plenary Session II

Time: Friday, December 13, 2019 - 9:00am-9:25am

Authors:

Bolan, N., University of Newcastle

Particulate plastic inputs to terrestrial and aquatic ecosystems cover range of particle size including microplastics and nanoplastics. Particulate plastics encompass a group of plastic polymers, and there are two types of particulate plastics that can enter the environment. Primary particulate plastics are manufactured and are a direct result of anthropogenic use of plastic-based materials. Secondary particulate plastics are plastic fragments resulting from the breakdown of larger plastic pieces. While wastewater discharge is a major source of particulate plastic input to aquatic ecosystem, application of biowastes such as biosolids and compost leads to particulate plastic input to soil. The UNEP identified that large quantities of particulate plastics found within the marine environment have originated from land-based sources. Both primary and secondary particulate plastics persist both in terrestrial (i.e., soil) and aquatic (i.e., marine and freshwater) ecosystems.

Particulate plastics input to soil can have both beneficial and detrimental impacts on soil characteristics and organisms. For example, polyacrylamide is used to promote flocculation and soil aggregation, thereby mitigating soil erosion. Particulate plastics impact environmental contaminant interactions in soil by serving as a vector for contaminant transport and by altering contaminant bioavailability in soil. As a vector, particulate plastics release toxic chemicals added during the manufacture of plastics to the soil environment. Despite the link between marine-based and to land-based sources of particulate plastic inputs, the majority of scientific research on particulate plastics has focused on their effects in aquatic environments. This presentation covers the sources, distribution and impacts of particulate plastics in soil.

The Need for Sustainable Management of Plastics and Microplastics- Chemical Toxicity and Health Impacts

Conference: Technical and Engineering Challenges of Addressing Sustainable Development

Year: 2019

Proceeding: Engineering Sustainable Development 2019

Group: Abstract Submissions

Session: Plenary Session II

Time: Friday, December 13, 2019 - 9:25am-9:50am

Authors:

Wong, M. H., The Education University of Hong Kong

A wide range of persistent and toxic chemicals is generated through the use of various consumer products: pharmaceuticals and personal care products (PCPPs), detergents, disinfectants, plasticizers, preservatives, microplastics, etc. These chemicals and their metabolites can be found in different environmental media, including wastewater, agricultural and urban runoff, rivers, ground and drinking water. The conventional wastewater and (drinking) water treatment processes are not always efficient in removing them. Some of these chemicals may enter the food chain, posing health threats, mainly through consumption of contaminated seafood. The major objectives of this presentation are to review the current status on the sources, fates and health effects of these chemicals, with emphasis on phthalates, bisphenol A, perfluorooctane sulfonate (PFOS), and PCPPs, citing some local examples in South China . It is also intended to review the exposure pathways of these chemicals, and their potential associations with body loadings and associated health effects. It is known that autism spectrum disorders in children is related to exposure to various environmental pollutants. These endocrine disrupting chemicals are widely detected in indoor dust, food items such as fish and meat, and blood plasma of local residents. There seems to be an urgent need to study the potential toxic effects of microplastics, including those derived from PCPPs, as they are now widely distributed in the coastal environment, and their uptake mechanisms by seafood, and health impacts on consumers are largely unknown. It seems essential to manage these emerging chemicals of concern more effectively, both nationally and internationally.

Invited Talk: Green and Sustainable Remediation of Contaminated Groundwater: Links to Sustainable Development Goals

Conference: Technical and Engineering Challenges of Addressing Sustainable Development

Year: 2019

Proceeding: Engineering Sustainable Development 2019

Group: Abstract Submissions

Session: Invited Session 4 - Sustainable Energy and Water Practices

Time: Friday, December 13, 2019 - 10:15am-10:40am

Authors:

Hou, D., Tsinghua University

In recent years, the concept of â€˜Green and Sustainable Remediationâ€™ (GSR) has emerged. It has provided a specific context for wiser decision making in groundwater remediation, with an emphasis placed on the overall Net Environmental Benefit (NEB) of remediation. GSR provides a framework where waste materials are granted greater value than virgin materials, which typically present greater cost in a life cycle. For the same reasons, *in situ* is valued above *ex situ* remediation, and low cost is valued above high cost remediation, so that the worldâ€™s limited resources can be put to best use. The advancement of GSR principles bring about an incentive to promote new remediation technologies, with the aim in mind of enhancing NEB. This departs from the traditional driving incentive, which, historically, has been to develop the best available technology (in terms of the reduction of risk from groundwater contaminants) not entailing excessive costs. As such, the growth of GSR in the remediation industry should proceed hand-in-hand with the production of new technologies. We further discuss how the GSR movement in the remediation industry is linking sustainable development goals (SDGs) for health (e.g., Goal 3) with socio-economic goals (e.g., Goal 8).

Invited Talk: Rational Design for Efficient and Stable Halide Perovskite Solar Cells

ENGINEERING
SUSTAINABLE
DEVELOPMENT

Conference: Technical and Engineering Challenges of Addressing Sustainable Development

Year: 2019

Proceeding: Engineering Sustainable Development 2019

Group: Abstract Submissions

Session: Invited Session 4 - Sustainable Energy and Water Practices

Time: Friday, December 13, 2019 - 10:40am-11:05am

Authors:

Noh, J. H., Korea Universdity

The haldie perovskite solar cells (HPSCs) have shown impressive progress in power conversion efficiency (PCE) in the last several years. This progress is attributed to developments of device architecture, perovskite halides, and fabrication process based on materials and devices engineering. However, there still remain many challenging issues on the HPSCs to be commercialized in the various applications in terms of PCE and long-term stability. For example, a breakthrough in materials and devices engineering is required for further improvement of PCE and long-term stability as well as the development of the scalable process. Among them, we have focused on the management of charge collection including a reduction of open circuit voltage deficit that is a difference between band-gap of the light absorbing halide and open-circuit voltage of the completed device as well as an increase in fill factor. The voltage deficit and fill factor are the result of a complex combination of factors including trap concentration, interfacial imperfection, energy band position arrangement, intrinsic electrical conduction, and even morphology of layers in perovskite solar cells. In this talk, our approaches will be introduced to improve charge collection properties and long-term stability in HPSCs in terms of n-type, p-type and light absorbing halide materials engineering.

Invited Talk: DOC Release from Peatlands By Reduction in Acid Deposition

ENGINEERING
SUSTAINABLE
DEVELOPMENT

Conference: Technical and Engineering Challenges of Addressing Sustainable Development

Year: 2019

Proceeding: Engineering Sustainable Development 2019

Group: Abstract Submissions

Session: Invited Session 4 - Sustainable Energy and Water Practices

Time: Friday, December 13, 2019 - 11:05am-11:30am

Authors:

Kang, H., Yonsei University

Increases in dissolved organic carbon (DOC) concentrations in streams have been recorded at locations across the world. A large proportion of DOC originates from watersheds with organic soils such as peatlands, which store over a third of global soil organic carbon. One suggested mechanism for the enhanced decomposition of peat and the releases of DOC is recovery from acidiï¬cation. However, no biological role in the process has yet been identiï¬ed. Here, we report extracellular enzyme activities and microbial composition in peatlands of Korea, the UK, Japan and Indonesia, and ï¬nd higher pH to promote phenol oxidase activities, greater abundances in Actinobacteria and fungi, and enhanced pore-water DOC concentrations. Our pH manipulation experiments also showed that increase in pH enhanced phenol oxidase activity and DOC production with greater Actinobacterial and fungal abundances. Finally, knockout or addition of phenol oxidase dramatically changed DOC and phenolic production, indicating the central role of phenol oxidase in DOC mobilization.

Alternative Green FUEL from Natural Waste Using DEEP Eutectic Solvent: Experiment and COSMO-RS MODEL

Conference: Technical and Engineering Challenges of Addressing Sustainable Development

Year: 2019

Proceeding: Engineering Sustainable Development 2019

Group: Abstract Submissions

Session: Parallel Session 5A - Sustainable Energy

Time: Friday, December 13, 2019 - 11:25am-11:50am

Authors:

Anantharaj, R., SSN College of Engineering, Rajiv Gandhi Salai (OMR), Kalavakkam, Tamilnadu-603110.
P, S., SSN College of Engineering
M, V., SSN College of Engineering
Vivek,, M. S., SSN College of Engineering, Rajiv Gandhi Salai (OMR), Kalavakkam, Tamilnadu-603110.
D, G., SSN College of Engineering

Diesel engines are the major source for transportation, thermal power generation and various miscellaneous industrial applications which is powered by non-renewable fossil fuels that powers most of the world. But due to fast depletion, increasing cost and emission of Smoke, Hydrocarbon, CO_X, NO_X and SO_X components during the combustion of fossil fuel, it is necessary to produce an alternative cleaner burning fuel. Biofuel, also known as green fuel is synthesized from natural sources such as plants and animals are renewable and abundant source of energy which is more environmental friendly than the conventional fossil fuels. In this work, Cardanol, which is extracted from CashewNut Shell Liquid (CNSL) is used as a primary source for the production of alternative green fuel. CNSL, a reddish brown viscous liquid is a mixture of unsaturated phenolic components namely anacardic acid, cardanol, cardol and 2-methyl cardol. Cardanol is extracted from CNSL by using a suitable Deep Eutectic Solvent (DES) and is then blended with different fuel blends such as Diesel, Petrol, Kerosene, Methanol and Ethanol at different proportions to increase its engine efficiency without any engine modifications. This newly synthesized biofuel will be characterized in terms of Total Fuel Consumption, Specific Fuel Consumption, Brake Power, Indicated Power, Mechanical Efficiency, Brake Thermal Efficiency, Indicated Thermal Efficiency and Emission of Smoke, Hydrocarbon, CO_X, NO_X and SO_X components

Attached Algae Cultivation for Coupling Sustainable Biomass Production and Environment Remediation

ENGINEERING
SUSTAINABLE
DEVELOPMENT

Conference: Technical and Engineering Challenges of Addressing Sustainable Development

Year: 2019

Proceeding: Engineering Sustainable Development 2019

Group: Abstract Submissions

Session: Parallel Session 5A - Sustainable Energy

Time: Friday, December 13, 2019 - 11:50am-12:15pm

Authors:

Kim, S., Sandia National Laboratories
Davis, R., Sandia National Laboratories
Monroe, E., Sandia National Laboratories

Algae has always been playing critical role in humanity by being very bottom of the food chain as well as responsible for more than half of oxygen generation on earth, further now is considered one of the most captivating alternative energy resources to replace fossil fuel. Algae has the potential to be used not only in biofuel, but also in many other products such as food supplements, animal feed, bioplastics, and cosmetics. This inexhaustible potential of algae has continued to draw attention to itself in academia and industry. Especially, efforts on the development of attached algae cultivation have been proliferated since it significantly reduces harvest/dewatering cost and contamination risk, which were the biggest challenges that typical suspended cultivation system had have. In this talk, Sandia National Labsâ€™ effort on attached algae cultivation for coupling sustainable biomass production and environment remediation will be discussed. Sandia has been deploying periphytic attached algae flow-ways, formerly known as Algal Turf Scrubber (ATS), at multiple contaminated location in US and running multiple years to see its sustainability in biomass productivity and ability to remove nutrients from contaminated source water. As a result, we were able to achieve consistent nutrient removal and biomass productivity without any crash for years at all our deployment locations. In addition, another type of attached cultivation technique, porous substrate-based attached cultivation will be introduced and its potential in algae-based high value product market will be discussed.

COSMO-RS Model Prediction of Polarity, Solubility, Activity Coefficient at Infinite Dilution, Hentry's Constant and Vapour Pressure of Hydrogen in Imidazolium Based Ionic Liquids at Different Temperatures (298.15 to 368.15)K

ENGINEERING
SUSTAINABLE
DEVELOPMENT

Conference: Technical and Engineering Challenges of Addressing Sustainable Development

Year: 2019

Proceeding: Engineering Sustainable Development 2019

Group: Abstract Submissions

Session: Parallel Session 5A - Sustainable Energy

Time: Friday, December 13, 2019 - 12:15pm-12:40pm

Authors:

Anantharaj, R., SSN College of Engineering, Rajiv Gandhi Salai (OMR), Kalavakkam, Tamilnadu-603110.
Kadambanathan, K., SSN College of Engineering, Rajiv Gandhi Salai (OMR), Kalavakkam, Tamilnadu-603110.
Vivek,, M. S., SSN College of Engineering, Rajiv Gandhi Salai (OMR), Kalavakkam, Tamilnadu-603110.

Hydrogen is an alternative fuel in the future due to avoiding global warming, global energy problems, and environmental impact and thereby to preserve the environment and to safeguard human health. On the other hand, the scarcity of petroleum or its reduced value, as a fuel with respect to non-energetic uses and Sevier environmental issues due to emission of CO_X,SO_X and NO_X during combustion of fossil fuel or liquid fuels. In this context, the use of ionic liquids (ILâ €™s) such as Imidazolium salt as a sustainable hydrogen storage material due to their unique properties such as no vapour pressure, high thermal stability, fast hydrogenation and dehydrogenation reaction at moderate temperature than that of lithium boron hydride ($LiBH_4$) and ammonia borane (NH_3BH_4). Therefore, in this work, the polarity, solubility, activity coefficient at infinite dilution and vapour pressure of hydrogen in Imidazolium based ionic liquids (alkyl range from methyl - decyl with limited fluorinated anions such as tetrafluoroborate (BF_4), hexafluoroborate (PF_6), triflate (TFO), trifluoroacetic acid (TFA) and bis (trifluoromethane) sulfonamide (NtF_2). This results were analyzed in terms of H-H interaction, charge â€“ charge interaction and electron pair affinity at atom and molecular level using COSMO-RS model. Because the specific interaction between ILâ€™s and hydrogen atom or molecules are locally important for nucleation and enhance the fast hydrogenation and dehydrogenation reaction at atmospheric or moderate condition.

Methyltriphenyl-Phosphoinum Bromide-Based Dess for Flue Gas Decarbonization: Experimental and Simulation Approach

ENGINEERING
SUSTAINABLE
DEVELOPMENT

Conference: Technical and Engineering Challenges of Addressing Sustainable Development

Year: 2019

Proceeding: Engineering Sustainable Development 2019

Group: Abstract Submissions

Session: Parallel Session 5A - Sustainable Energy

Time: Friday, December 13, 2019 - 12:40pm-1:05pm

Authors:

Haider, M. B., Rajiv Gandhi Institute of Petroleum Technology
Tripathi, M. M., Rajiv Gandhi Institute of Petroleum Technology
Kumar, R., Rajiv Gandhi Institute of Petroleum Technology

In this work three different methyltriphenylphosphonium bromide based deep eutectic solvents (DESs) were synthesized and their ability to capture CO_2 was experimentally investigated. Ethylene glycol, diethylene glycol and glycerol were used as hydrogen bond donor for DESs synthesis. The CO_2 uptake in DESs were measured using pressure drop method at different temperature and pressure. Critical properties of synthesized DESs were calculated using â€œModified VFTâ€ equation. Peng-Robinson equation of state was used to model the experimental vapor liquid equilibria data of CO_2/DESs system. In addition, flue gas decarbonization process was simulated using predicted properties data and experimental VLE data. Sensivity analysis was employed to the DESs-based process to obtain the optimal parameters. In addition, heat integration was used to further lower down the energy requirement.

Oscillating Flow Fields in Thermoacoustic Energy Systems

ENGINEERING SUSTAINABLE DEVELOPMENT

Conference: Technical and Engineering Challenges of Addressing Sustainable Development

Year: 2019

Proceeding: Engineering Sustainable Development 2019

Group: Abstract Submissions

Session: Parallel Session 5A - Sustainable Energy

Time: Friday, December 13, 2019 - 1:05pm-1:30pm

Authors:

Harikumar, G., Nanyang Technological University
Shen, L., Nanyang Technological University
Wang, K., Zhejiang University
Dubey, S., Nanyang Technological University
Duan, F., Nanyang Technological University

A thermoacoustic Stirling engine can produce electric power by simultaneously utilizing the cold energy from liquefied natural gas re-gasification process and the waste heat from the exhaust of gas turbines using a pair of linear alternators. However, the actual efficiency of the device when tested was much lower than that predicted by the design software. This might be due to un-accounted non-linear effects like entrance effects, vortex shedding, acoustic streaming etc. which affect the energy transfer processes occurring in the vicinity of the thermoacoustic core or stack/regenerator. It is therefore necessary to further understand and investigate these non-linear effects in an oscillating flow in order to develop more accurate models to predict the performance of thermoacoustic systems and improve the efficiency. To study and observe these phenomenon, a standing wave thermoacoustic system was setup with a parallel plate stack made of acrylic for its simplicity. A loudspeaker is used to generate pressure oscillations inside the standing wave resonator and create a temperature gradient across the stack ends. The flow near the thermoacoustic stack is visualized using particle image velocimetry technique. We can observe vortex generation at the edges of the plates followed by elongation and shredding as the cycle progresses. These vortices interact with the stack and each other in reverse flow to form complex structures that might affect the heat transfer process and the efficiency of the system.

Recovery of the Waste Heat from Low Temperature and High Temperature Proton Exchange Membrane Fuel Cell Using Water Based Adsorption Chiller

Conference: Technical and Engineering Challenges of Addressing Sustainable Development

Year: 2019

Proceeding: Engineering Sustainable Development 2019

Group: Abstract Submissions

Session: Parallel Session 5A - Sustainable Energy

Time: Friday, December 13, 2019 - 1:30pm-1:55pm

Authors:

Naseem, M., Dongguk University
Sagheer, M. U., Dongguk University
Song, M., Dongguk University
Lee, S., Dongguk University
Park, S., Dongguk University

With the immense problem of global warming, clean energy resources are the way to salvation for the future generations. Fuel cell power plants are highly promoted to serve this as they give off zero emissions making them environment friendly. Low temperature Proton Exchange Membrane fuel cells are utilized for generating electrical energy from hydrogen gas mostly used in combined heat and power cogeneration for maximum efficiency. For a sustainable energy model, the concept of trigeneration is introduced by various researchers in which cooling is also achieved from the waste heat of the fuel cell. Moreover for an environment friendly system, adsorption chilling is the most favorable as it does not involve any chlorofluorocarbons nor any other toxins as water is the working fluid for this device. Regeneration of the adsorbent consumes waste heat from the fuel cell. The effect of using a low and/or a high temperature Fuel cell is studied when used in a trigeneration system, and it is found that despite lower efficiency of high temperature fuel, the net system efficiency would be maximum. In contrast to low temperature fuel cell, which is optimal for use in a combined heat and power cogeneration system, high temperature fuel cell serves the best cause when used in a trigeneration system. This is mainly due to the rapid drop in performance of the adsorption chiller with decreasing adsorbent regeneration temperature.

Using Neural Network to Predict Pyrolysis Biochar Production and Its Impact on Low-Carbon Agricultural Development in Scotland

ENGINEERING
SUSTAINABLE
DEVELOPMENT

Conference: Technical and Engineering Challenges of Addressing Sustainable Development

Year: 2019

Proceeding: Engineering Sustainable Development 2019

Group: Abstract Submissions

Session: Parallel Session 5B - Economic, Social, Climatic and Environmental Issues

Time: Friday, December 13, 2019 - 11:25am-11:50am

Authors:

Li, Y., University of Glasgow
You, S., University of Glasgow

Sustainable agricultural development is one of the major challenges that call for effective actions. Agricultural residues have been long used as an important bioresource for energy and material recovery. In recent, significant attention has been paid to the production of biochar from the pyrolysis of agricultural waste, and the application biochar for soil amendment and carbon abatement.

Based on an extensive review of existing data, this study developed a framework consisting of artificial neural networks (ANN) and the least squares support vector machine (LS-SVM) modelling to predict the biochar yield and quality from the pyrolysis of agricultural waste biomass and the impacts of biochar soil application on crop productivity. The framework utilised the adaptive neuro-fuzzy inference system approach and is able to predict biochar production under different pyrolysis conditions such as feedstock types (compositions), temperature and heating rates. The framework was used to evaluate the carbon abatement and agriculture productivity enhancement potential of soil amendment in Scotland by biochar produced from major Scottish agricultural residues

Materials Conservation and Circularity Index Applied to Ccus

Conference: Technical and Engineering Challenges of Addressing Sustainable Development

Year: 2019

Proceeding: Engineering Sustainable Development 2019

Group: Abstract Submissions

Session: Parallel Session 5B - Economic, Social, Climatic and Environmental Issues

Time: Friday, December 13, 2019 - 11:50am-12:15pm

Authors:

Naraharisetti, P. K., Institute of Chemical & Engineering Sciences

Several indices for circularity index have been proposed and applied to various organizations in the manufacturing sectors; however, the application of circularity index to the Chemical Process Industries and CCUS (Carbon Capture, Utilization Storage) is rather limited. Nevertheless, circular economy and CO_2 capture are considered as the means of mitigating climate change associated with the release of CO_2 and it is important that we apply an appropriate Circularity Index to this sector. In view of this, we consider the Materials Circularity Index (MCI) proposed by the Ellen MacArthur foundation and modify it to develop the modified Materials Circularity Index (mMCI). We then apply the mMCI on the process of CO_2 mineralization and compare it with the processes for Carbon Utilization. In addition, we propose that all linear processes are not the same and that it is important to consider the mMCI along with an index for resource use and conservation (the Resource Conservation Index, RCI) so that the different linear (poorly designed â€˜Circular Economyâ€™) processes and circular processes can be compared against one another. We further show that both these indices can be combined and call the same as the â€˜Materials Conservation and Circularity Indexâ€™, MCCI. For the purpose of this work, we analyse three case studies and compare one against the other. Briefly, one case study on CO_2 mineralization consisting of three scenarios and two cases of CO_2 utilization are considered.

Immobilization of Potentially Toxic Elements in Contaminated Soils Using Various Soil Amendments

ENGINEERING SUSTAINABLE DEVELOPMENT

Conference: Technical and Engineering Challenges of Addressing Sustainable Development

Year: 2019

Proceeding: Engineering Sustainable Development 2019

Group: Abstract Submissions

Session: Parallel Session 5B - Economic, Social, Climatic and Environmental Issues

Time: Friday, December 13, 2019 - 12:15pm-12:40pm

Authors:

Palansooriya, K. N., Korea University
Ok, Y. S., Korea University

Soil contamination by potentially toxic elements (PTEs) has led to adverse environmental impacts. In this review, we discussed remediation of PTEs contaminated soils through immobilization techniques using different soil amendments with respect to type of element, soil, and amendment, immobilization efficiency, underlying mechanisms, and field applicability. Soil amendments such as manure, compost, biochar, clay minerals, phosphate compounds, coal fly ash, and liming materials are widely used as immobilizing agents for PTEs. Among these soil amendments, biochar has attracted increased interest over the past few years because of its promising surface properties. Integrated application of appropriate amendments is also recommended to maximize their use efficiency. These amendments can reduce PTE bioavailability in soils through diverse mechanisms such as precipitation, complexation, redox reactions, ion exchange, and electrostatic interaction. However, soil properties, such as soil pH, and clay, sesquioxides and organic matter content, and processes, such as sorption/desorption, and redox processes, are the key governing factors determining amendmentsâ€™ efficacy for PTEs immobilizing soils. Selecting proper immobilizing agents can yield cost-effective remediation techniques and fulfill green and sustainable remediation principles. Furthermore, long-term stability of immobilized PTE compounds and the environmental impacts and cost effectiveness of the amendments should be considered before application.

Development of Low Cost Deep Eutectic Solvents for Extraction of Polyphenols from Natural Waste

ENGINEERING SUSTAINABLE DEVELOPMENT

Conference: Technical and Engineering Challenges of Addressing Sustainable Development

Year: 2019

Proceeding: Engineering Sustainable Development 2019

Group: Abstract Submissions

Session: Parallel Session 5B - Economic, Social, Climatic and Environmental Issues

Time: Friday, December 13, 2019 - 12:40pm-1:05pm

Authors:

Anantharaj, R., SSN College of Engineering, Rajiv Gandhi Salai (OMR), Kalavakkam, Tamilnadu-603110.
M, V., SSN College of Engineering
P, S., SSN College of Engineering
Vivek, M. S., SSN College of Engineering

Cashew Nut Shell Liquid (CNSL) is a low-cost material for unsaturated phenol, which in terms a useful by product of cashew industry. The nuts has a shell of about 1/ 8 inch thickness inside which is a soft honey comb structure containing a dark reddish brown viscous liquid. These liquid contains phenolic compounds such as cardanol, anacardic acid, 2-methyl cardol. The first two compounds are monohydric phenol and others two compounds are dihydric phenol. The thoughtful process of these work includes extraction of phenolic compounds using low cost deep eutectic solvents (DESs). The extracted sample will be analyzed by UV, FTIR and NMR analysis for screening the potential solvent and generate phase equilibria diagram using the tie line composition for better understanding the solute-solvent interaction and determine the efficiency of the separation process at atmospheric conditions. The reliability and consistency of the experimental data will be evaluated by using Othmer-Tobias and Hand equation. The RMSD value are calculated for the predicted COSMO-RS tie-line data and experiment data. The effectiveness of extraction of cardanol using DESs are studied by COSMO-RS predictions along with the experimental LLE data with selectivity, distribution coefficient and performance index. The DFT method will be used to produce more reliable geometries for hydrogen-bond formation in the mixture of solvent and Phenolic compounds. On the other hand, the feasibility test was carried out for solvent regeneration and reuse which makes water insensitive solvent attractive process involving multiple cycles.

Molecular Dynamics Simulations on Extractive Desulfurization of Fuels By Cobalt Chloride/Choline Chloride/Peg-200 Metal Based Deep Eutectic Solvents

ENGINEERING SUSTAINABLE DEVELOPMENT

Conference: Technical and Engineering Challenges of Addressing Sustainable Development

Year: 2019

Proceeding: Engineering Sustainable Development 2019

Group: Abstract Submissions

Session: Parallel Session 5B - Economic, Social, Climatic and Environmental Issues

Time: Friday, December 13, 2019 - 1:05pm-1:30pm

Authors:

Kukpayev, A., Nazarbayev University

Considering environmental issues, desulfurization process is of paramount importance to produce fuel with low sulfur content. In recent decades, novel sustainable deep eutectic solvents (DESs) illustrated high extractive capacity and efficiency in such desulfurization processes. In this work, interactions of metal-based DESs with model oil consisting of thiophenic compounds were analyzed by using molecular dynamic simulation. In particular, a combination of choline chloride (ChCl): polyethylene glycol (PEG-200): cobalt chloride ($CoCl_2$)at a ratio 1:2:1 and a combination of tetrabutylammonium chloride (TBAC): PEG-200: $CoCl_2$ at a ratio 1:2:0.25 were simulated, separately, with model oil at 25°C & 1bar. The simulations reveal lower interaction energies between the constituent compounds of MDESs as compared with their pure state supporting the formation of MDES. Furthermore, detailed analysis of the extraction process reveals that both MDES have strong interactions with refractory sulfur compounds in model oil indicating a high desulfurization efficiency. While extraction efficiency weakly correlates with variation in the hydrogen bond acceptor, hydrogen bond donor, it strongly depends on temperature. As DESs are highly viscous, we herein propose a trade-off between adding water to decrease DESâ€™s viscosity and the efficiency of desulfurization. Taken together, the presented herein present mechanism of desulfurization at molecular level and develop guidelines for practical usage of MDESs in the process.

Invited Talk:A Building Momentum for Global Sustainability Using the Principles of Behavioral Economics

ENGINEERING
SUSTAINABLE
DEVELOPMENT

Conference: Technical and Engineering Challenges of Addressing Sustainable Development

Year: 2019

Proceeding: Engineering Sustainable Development 2019

Group: Abstract Submissions

Session: Invited Session 6 - Economic, Social, Climatic and Environmental Issues

Time: Friday, December 13, 2019 - 1:25pm-1:50pm

Authors:

Seay, J. R., University of Kentucky

Despite the current threat from climate change, plastic collecting in the worldâ€™s oceans, and the steady loss of biodiversity, the world continually fails to take action with regard to our rapidly changing ecosystem. Unfortunately, waiting on governments to act is no longer a viable option. Rapid change is needed and the pace of diplomacy is simply too slow. Democratic governments are reactionary and due to the principle of hyperbolic discounting, taking action to solve future problems is not a priority, even as the threat of potential ecological catastrophe draws ever closer. Change is in the hands of individuals, and it is our decisions and behaviors that will influence the future of our planet and our ability to inhabit it. Therefore, building momentum for sustainable behavior must begin with individuals. The neoliberal approach to environmental protection posits that individuals are motivated by rational self-interest, and that economic incentives are necessary to achieve environmental goals. However, recent research suggests that monetary gain alone actually negatively impacts behavior, and often neglects the rural poor. As a result, models for projects designed to benefit the environment need more than just a monetary incentive, they must incorporate all three pillars of sustainability: environment, economy and society. This contribution will describe the role behavioral economics plays in the choices made by producers and consumers.

Invited Talk: Advanced Membranes for Desalination and Water Treatment

ENGINEERING SUSTAINABLE DEVELOPMENT

Conference: Technical and Engineering Challenges of Addressing Sustainable Development

Year: 2019

Proceeding: Engineering Sustainable Development 2019

Group: Abstract Submissions

Session: Invited Session 6 - Economic, Social, Climatic and Environmental Issues

Time: Friday, December 13, 2019 - 1:50pm-2:15pm

Authors:

Lee, J. H., Korea University
Jeon, S., Korea University
Park, S. J., Korea University

Polyamide (PA) thin film composite (TFC) membranes, consisting of an ultrathin PA selective layer on a porous support, are the predominant material platform of membranes for desalination and water treatment. The PA selective layer is conventionally prepared via interfacial polymerization of multifunctional amine and acyl chloride monomers. Despite of its high permselectivity, the PA layer has a very complex and rough structure, which limits the performance and durability enhancement of the membrane and hampers our understanding of the structure-property relationship of the membrane. This concern necessitated the development of the fabrication technique of the PA layer with the precisely controlled and well-defined structure. In this talk, advanced membrane materials with well-defined physical and chemical PA structures were fabricated via new fabrication methods, including layered interfacial polymerization and dual slot coating, and new building block chemistry. The prepared PA layers were used as a model material platform for studying its structure-property relationship as well as a practical material platform for improving the membrane performance and durability.

Invited Talk: Geopolymerisation Using Coal Fly Ash and Wood Fly Ash with Internal and External Alkali Sources

Conference: Technical and Engineering Challenges of Addressing Sustainable Development

Year: 2019

Proceeding: Engineering Sustainable Development 2019

Group: Abstract Submissions

Session: Invited Session 6 - Economic, Social, Climatic and Environmental Issues

Time: Friday, December 13, 2019 - 2:15pm-2:40pm

Authors:

Wang, C. H., National University of Singapore
Prabhakar, A. K., National University of Singapore

Geopolymers offer several advantages over Ordinary Portland cement(OPC), especially in terms of environmental sustainability and physiochemical properties. Here geopolymers were formulated using coal fly ash with high iron oxide and calcium oxide contents and displayed improved heat resistance, but a relatively lowered acid resistance than conventional OPC. The costs of alkaline reagents used and the benefit of geopolymer selling were the major factors affecting the economic feasibility of geopolymer production. To further lower this cost incurred due to use of external alkaline reagents and to facilitate the recycling of more ash, a follow-up study using wood fly ash as the alternative source of alkaline activator for the geopolymerisation with coal fly ash was done. However, the leaching of heavy metals from the wood fly ash could potentially pose an environmental concern. Therefore, the focus of this study was to reduce the leachability of wood fly ash throughout the geopolymerisation process and to understand the effects of various factors on the leaching process. Our result showed that the dominant factor was the type of alkaline activator. By comparing the leaching data to the Denmarkâ€™s leaching criteria, the best performing geopolymer sample uses wood fly ash as its alkaline activator, thus doing away with any external alkaline source and reducing the overall cost.

Invited Talk: Biochar As a Support for Triumphing Sustainable Development Goals

ENGINEERING SUSTAINABLE DEVELOPMENT

Conference: Technical and Engineering Challenges of Addressing Sustainable Development

Year: 2019

Proceeding: Engineering Sustainable Development 2019

Group: Abstract Submissions

Session: Invited Session 6 - Economic, Social, Climatic and Environmental Issues

Time: Friday, December 13, 2019 - 2:40pm-3:05pm

Authors:

Vithanage, M., Ecosphere Resilience Research Center, Faculty of Applied Sciences, University of Sri Jayewardenepura

Biochar is a carbonaceous material produced under pyrolysis or gasification of various waste biomasses as feedstocks with lignin, cellulose and hemicellulose. Composition of biochar plays a role in determining its application. Biochar has received strong attention from researchers in diverse fields due to its unique and universal characteristics. The Global Goals (17 in number) are set up to end poverty, protect the planet and ensure that all people enjoy peace and prosperity known as Sustainable Development Goals (SDGs) became an universal action. The objective here is to assess biocharâ€™s capacity in achieving the said SDGs. Biochar from agri-waste amalgamation with wastewaters may reduce the fertilizer use will reduce poverty while soil conditioning with biochar improves soil fertility converting poor soils into productive land will leads to zero hunger. Biochar is extensively used in soil and water remediation and hence achieving 3rd and 6th goals; good health & well-being and clean water sanitation are promising. Waste to energy helps in waste minimization via producing biochar, which can be used to remediate environmental pollution from landfills. Converting waste into usable material leads to circular economy with material recycling demonstrate sustainable production and consumption. Sri Lanka is trying to generate biochar using municipal solid waste as a multi solution approach to open dumping of waste. Biochar reduces green-house gases while its addition to water and soil reduce nutrient leaching helps life below water. Reclamation, remediation, restoration improve life on land. Hence, biochar is the future material for the globe to achieve sustainability.

A Sustainable Approach Towards Non-Photosynthetic Conversion of Carbon-Di Oxide to Bio-Diesel Utilizing Mixed Microbial Culture-Steps Towards Climate Change

ENGINEERING
SUSTAINABLE
DEVELOPMENT

Conference: Technical and Engineering Challenges of Addressing Sustainable Development

Year: 2019

Proceeding: Engineering Sustainable Development 2019

Group: Abstract Submissions

Session: Session 7 - Economic, Social, Climatic and Environmental Issues

Time: Friday, December 13, 2019 - 3:00pm-3:25pm

Authors:

Raghuvanshi, S., Birla Institute of Technology and Science (BITS) - BITS Pilani, Pilani Campus, Rajasthan, India
Gupta, S., Birla Institute of Technology & Science, Pilani
Mishra, S., IIT Delhi, India

Concentration level of carbon dioxide (CO_2) in the atmosphere has reached around 400 parts per million, in the last three million years. This has resulted in the problem of global warming and climate change. One of the efficient solution which is suggested is the carbon capture utilization using biological route. Hence, in the present abstract, we are discussing our experimental work (batch mode), which involves removal of CO_2 utilizing mixed microbial culture obtained from city`s sewage treatment plant (STP). This can be referred as an example of non-photosynthetic conversion of CO_2. The work discusses few sets of experiments which were carried out to analyze the leachates (biomass + supernatant) obtained from experiments using FTIR and GC-MS for the identification of value-added products. The feasibility of the process was checked, also, theoretically by performing approximate carbon balance and thermodynamic assessment. The mixed microbial culture has shown the removal efficiency of around 78% at 100 ppm of Fe(II) concentration which is more when results were compared with individual strains of*P.putida (RE â€“ 67.44%)* and *E.cloacae* (RE â€“ 58.44%). The by-product characterization confirmed the presence of fatty acids (C7â€“C25) and hydrocarbons (C9â€“C32) in the extractable amount. The FAME content yield obtained from fatty acids, was found to be at its maximum of 91.55% for the mixed microbial culture and qualified as biodiesel. The present work paved a possible pathway for non-photosynthetic production of biodiesel, which is a step towards sustainable processes and has a potential to fulfill present day industrial needs.

Highly Active and Stable Bimetallic Rhfe/HAP Catalyst for Direct and Non-Thermal Plasma Synergistic Catalysis of N2o Decomposition

ENGINEERING
SUSTAINABLE
DEVELOPMENT

Conference: Technical and Engineering Challenges of Addressing Sustainable Development

Year: 2019

Proceeding: Engineering Sustainable Development 2019

Group: Abstract Submissions

Session: Session 7 - Economic, Social, Climatic and Environmental Issues

Time: Friday, December 13, 2019 - 3:25pm-3:50pm

Authors:

Tan, X., School of Chemical Engineering and Technology
Qi, S., School of Chemical Engineering and Technology
Lu, Q., School of Chemical Engineering and Technology
Shi, L., School of Chemical Engineering and Technology

The motivation for this paper is to further improve the efficiency of N_2O decomposition on bimetallic catalysts which often possess a superior performance than their monometallic counterparts by non-thermal plasma (NTP) synergistic effect. The N_2O decomposition in the presence of hydroxylapiatite (HAP) supported Rh and Fe monometallic and bimetallic catalysts have been investigated. Also NTP synergistic effect has been inspected. The HAP is prepared at different pH values and the Rh/Fe supported catalysts synthesized by impregnation method. The research illustrates that basic sites are in favor of N_2O decomposition. The bimetallic catalyst RhFe/HAP-11 possesses a superior performance compared to its monometallic counterparts. The T_{50} values of RhFe/HAP-11 is 285°C and that of the catalyst with hydrogen pretreatment is 207°C which is the minimum value as far as we know. When plasma is applied to synergistic catalysis, N_2O conversion on RhFe/HAP-11 at 200°C increases a lot from 35.7% to 84.1% (one-stage mode) and 81.0% (two-stage mode). And the N_2O conversion increases with the rise of plasma discharge power. The results of the catalyst characterization by XPS, SEM, TEM demonstrated that metal particles on the catalyst without pretreatment are in oxidation state and uniformly dispersing, zero-valence metal on catalyst make the N_2O conversion higher. TG curve and continue test illustrate that the catalyst has ideal thermal stability. All the results show that with RhFe/HAP-11 catalyst and non-thermal plasma synergistic catalysis in the N_2O decomposition, ideal conversion can be gotten in the low temperature range.

High-Value Added Valorization of Food Waste Streams to Rhamnolipids

ENGINEERING SUSTAINABLE DEVELOPMENT

Conference: Technical and Engineering Challenges of Addressing Sustainable Development

Year: 2019

Proceeding: Engineering Sustainable Development 2019

Group: Abstract Submissions

Session: Session 7 - Economic, Social, Climatic and Environmental Issues

Time: Friday, December 13, 2019 - 3:50pm-4:15pm

Authors:

Kaur, G., Hong Kong Baptist University
Wong, J., Hong Kong Baptist University
Uisan, K., Hong Kong Baptist University
Patria, R. D., Hong Kong Baptist University
Davidraj, J., Hong Kong Baptist University

Globally, food waste constitutes 30-50% of total municipal solid waste while landfilling remains the main treatment technology. From a circular economy perspective, this treatment option is not sustainable due to greenhouse gases emission and breakdown of nutrient cycling. Meanwhile, the recent sustainable development goals emphasize on food security, environmental protection and material and energy efficiency, and therefore, have become the key drivers for food waste valorization to high-value products. In the present work, we investigated the feasibility of bioconversion of mixed, post-consumption leftover food waste into rhamnolipids biosurfactant using *Pseudomonas aeruginosa*. Food waste was obtained from university cafeteria in Hong Kong. The food waste hydrolysate was obtained by mixed culture acidogenic fermentation instead of using expensive commercial enzymes. Preliminary experiments compared the rhamnolipids production efficiency on (food) waste streams including waste cooking oil and mixed food waste with refined hydrophilic substrates such as glucose and glycerol, and hydrophobic carbon sources such as oleic acid and fresh canola oil. Food waste emerged as the best feedstock for rhamnolipids production and was subsequently used for fermentation in a laboratory bioreactor. Characterization of produced rhamnolipids was performed by Liquid Chromatography with tandem Mass Spectrometry (LC-MS/MS) and Fourier Transform Infrared Spectroscopy (FTIR). It was shown that food waste-derived rhamnolipids predominantly contained Rha-C10-C10 and Rha-Rha-C10-C12 congeners. Further improvement in product concentration and yield through optimization of fermentation conditions and design of various nutrient feeding fed-batch strategies shall be the topic of future investigation.

Sorption and Transport of Antibiotic Ciprofloxacin in Soils Amended with Colloidal Biochar and Halloysite

ENGINEERING SUSTAINABLE DEVELOPMENT

Conference: Technical and Engineering Challenges of Addressing Sustainable Development

Year: 2019

Proceeding: Engineering Sustainable Development 2019

Group: Abstract Submissions

Session: Session 7 - Economic, Social, Climatic and Environmental Issues

Time: Friday, December 13, 2019 - 4:15pm-4:40pm

Authors:

Ashiq, A., Ecosphere Resilience Research Center
Vithanage, M., Ecosphere Resilience Research Center, Faculty of Applied Sciences, University of Sri Jayewardenepura
Sarkar, B., Department of Animal and Plant Sciences, The University of Sheffield
OK, Y., Korea University

Fluoroquinolones are one of the most frequently used antibiotics in the veterinary industry which shows tremendous mobility in soils after manure application. Colloidal biochar (CBC) amendments in soils facilitate the immobility and enhance physicochemical properties to retain contaminants in soils. Objectives of this study were to characterize the CBC halloysite composite, further to determine the sorption of ciprofloxacin (CPX), to evaluate the effect on CPX retainment when soil amended with CBC and halloysites. Biochar produced from *Gliricidia sepium*were taken to produce the CBC using ball-miller grinder and characterized using particle analyser and scanning electron microscope (SEM). The abilities to adsorb CPX at different pH in soil to retain the CPX were examined using batch sorption experiments and column studies. The particles analysed were in the range of 350- 400 nm and showed encapsulation of halloysite nanotubes within the biochar through SEM. Results showed a 40% increase in the CPX retention in soil with 2% amendments of CBC with halloysite and showed optimum adsorption at pH 5. Higher adsorption performance in the acidic pH and in the zwitterionic state of CPX could be attributed to both p-pelectron donor-acceptor interaction arising from CBC and CPX along with cation exchange and electrostatic interaction from the halloysite and CPX. Overall, CBC composites with halloysites could be utilised as an effective amendments to immobilise antibiotics from contaminated soils.

A Study of the Physical Properties of Plastic Derived Fuel Oil Produced from Waste Plastic

ENGINEERING
SUSTAINABLE
DEVELOPMENT

Conference: Technical and Engineering Challenges of Addressing Sustainable Development

Year: 2019

Proceeding: Engineering Sustainable Development 2019

Group: Poster Submissions

Session: Poster Session

Authors:

Browning, S., University of Kentucky College of Engineering

Quigley, B., University of Kentucky
Kizza, R., Makerere University
Joshi, C., University of Kentucky College of Engineering
Seay, J. R., University of Kentucky

Plastic waste around the globe has become a massively detrimental issue that can lead to health complications and environmental destruction. Studies and prior experimentation suggest that thermally decomposing plastic will cause the carbon bonds to break apart and form a fuel oil similar in composition to diesel. Using a low-cost and simple processor called the Trash-to-Tank (3T) processor developed by the University of Kentucky Appropriate Technology and Sustainability (UKATS) Research Team, waste plastic can be converted into fuel, known as Plastic Derived Fuel Oil (PDFO) and sold as a diesel alternative.

The 3T processor has been specifically designed to be appropriate for underdeveloped regions where access to sophisticated literature are often lacking. Currently, the UKATS Research Team is conducting a pilot study in Uganda to assess the performance of the 3T process. In order to ensure that the fuel meet all appropriate regulatory requirements for use as a motor vehicle fuel in Uganda, emissions testing has been commissioned from the Ugandan Bureau of Standards. The laboratory results for the PDFO will be presented and compared with traditional number 2 fuel oil. The 3T process has been implemented in Uganda and has the potential to provide a low cost way for rural communities in developing regions to take ownership of the waste plastic problem in their communities. This process is part of the establishment of locally managed decentralized circular economies in underdeveloped regions and further can give women and other people living in poorer regions the ability to become entrepreneurs.

Catalytic Gasification of Plastics Using Ni Loaded Catalysts

ENGINEERING
SUSTAINABLE
DEVELOPMENT

Conference: Technical and Engineering Challenges of Addressing Sustainable Development

Year: 2019

Proceeding: Engineering Sustainable Development 2019

Group: Poster Submissions

Session: Poster Session

Authors:

Park, Y. K., University of Seoul

The activity of Ni/Si with the molar ratio of 0.5 and 1 and Ni/silica were tested on biomass gasification. The catalyst with the higher nickel content (molar ratio 0.5) for Ni/silicate and Ni/silica resulted in higher gas yield and more H_2 and CO generation while Ni/Si (0.5) generated less CO_2 compared to Ni/silica. The higher nickel content was used for the experiments to test silicate and silica as a support. The catalysts were characterized using BET and XRD analysis. Both the supports (silica and silicate) showed promising stability at higher temperature and can be employed in gasification

Cationic Network Polymers: The Impact of Counter-Anions and Porosities on for Atmospheric CO2 Fixation and Cyclic Carbonate Formation

ENGINEERING
SUSTAINABLE
DEVELOPMENT

Conference: Technical and Engineering Challenges of Addressing Sustainable Development

Year: 2019

Proceeding: Engineering Sustainable Development 2019

Group: Poster Submissions

Session: Poster Session

Authors:

Lim, W., KAIST
Yavuz, C. T., UCSB
Kim, D., KAIST
Garai, M., KAIST
Jeong, U., KAIST
Song, Y., KAIST

Non-redox carbon dioxide utilization through cycloaddition of CO_2 to epoxides offer great promise, but suffer from suitable heterogeneous catalysts are vital for their widespread implementation. We recently reported a highly efficient heterogeneous catalyst for CO_2 addition to epoxides based on a newly identified active catalytic pocket consisting of imidazolinium moieties, but there is a clear need for industrial implementation through systematic optimization. Different counter-anions and positively charged imidazolinium units in their backbones was prepared for optimizing in quantitative yield to cyclic carbonate. Catalytic actives sites derived from the counter-anions were characterized with heat of adsorption (Q_{st}) to identify CO_2 affinities and less-reactive substrates to compare catalytic activities. Also, silica template produced increased porosity in catalysts that enhanced diffusion in porous networks. The systematic optimization of counter-anions and porosity with catalytic activities provides attractive new paths for further scale-up and industrial implementation.

CO2-Derived Porous Carbon-Metal Oxide Composite As a Lithium-Ion Battery Anode Material

ENGINEERING SUSTAINABLE DEVELOPMENT

Conference: Technical and Engineering Challenges of Addressing Sustainable Development

Year: 2019

Proceeding: Engineering Sustainable Development 2019

Group: Poster Submissions

Session: Poster Session

Authors:

Choi, W. Y., KAIST
Lee, J. W., Korea Advanced Institute of Science & Technology (KAIST)
Choi, J. W., Korea Advanced Institute of Science and Technology
Kim, H. T., Korea Advanced Institute of Science and Technology
Lee, D. K., Korea Advanced Institute of Science and Technology

As energy-related issues have emerged, lithium-ion batteries (LIBs) are widely commercialized in various electrical devices as environmentally friendly energy storage systems. Among anode materials in LIBs, metal oxides are studied as promising candidates because of their high theoretical capacity. However, some drawbacks of low electrical conductivity and volume changes bring limits to them to be used as anode materials. Thus, this study synthesized the composite of metal oxide with porous carbon to compensate the drawbacks of metal oxides. In the synthesizing process, carbon dioxide (CO_2), a representative greenhouse gas, was actively utilized in synthesizing the both porous carbon and metal oxide simultaneously. As a result, the porous carbon-metal oxide composite showed superior electrochemical performance when applied as an anode material, maintaining high capacities (~1100 mAh/g) over 300 cycles stably.

Effect of Biochar Pyrolysis Temperature on Chemical Immobilization of Pb and As in Contaminated Soils

ENGINEERING SUSTAINABLE DEVELOPMENT

Conference: Technical and Engineering Challenges of Addressing Sustainable Development

Year: 2019

Proceeding: Engineering Sustainable Development 2019

Group: Poster Submissions

Session: Poster Session

Authors:

Shin, H., Korea University
Dissanayake, P. D., Korea University
Igalavithana, A. D., Korea University
Yu, S., Sungkyunkwan University
Ryu, C., Sungkyunkwan University
Ok, Y. S., Korea University
Park, J., Sungkyunkwan University
Palansooriya, K. N., Korea University

Lead (Pb) and Arsenic (As) are major potentially toxic elements (PTEs) causing serious effects on soils and living organisms. Hence, the immobilization of PTEs is essential to maintain soil health. There are a variety of methods to immobilize PTEs in soils; and remediation using biochar soil amendment has been newly proposed. Objective of this research was to assess the effect of pyrolysis temperature of biochar for the immobilization of Pb and As in contaminated soils. The biochar were are produced from wood pellet with highly porous structure at four different pyrolysis temperatures, 400 °C, 500 °C, 600 °C, and 700 °C. The ability of the biochar to immobilize Pb and As in contaminated soils was investigated through a laboratory incubation for 30 days. Soil samples were analyzed for chemical and microbial properties such as pH, electrical conductivity, cation exchange capability, labile organic carbon, total C and N, and soil enzymes. The most suitable pyrolysis temperature will be found from the results to immobilize As and Pb contaminated soil.

Enhancing the Dehydrogenation Efficiency of Heterocyclic Aromatic Hydrocarbon By Pd and Pt-Based Catalysts

ENGINEERING
SUSTAINABLE
DEVELOPMENT

Conference: Technical and Engineering Challenges of Addressing Sustainable Development

Year: 2019

Proceeding: Engineering Sustainable Development 2019

Group: Poster Submissions

Session: Poster Session

Authors:

Gong, X., Xi'an Jiaotong University
Jiang, Z., Xi□an Jiaotong University

In recent decades, hydrogen energy has played an important role on the sustainable development of the word. It is necessary to provide an appropriate technology for hydrogen storage to promote its commercial applications. Previous reports indicated that liquid organic hydrogen carriers (LOHC) exhibited significant advantages in stationary hydrogen storage system. The N-ethylcarbazole (NECZ) and dodecahydro-N-ethylcarbazole (12H-NECZ) system has been regarded as an ideal candidate of LOHC. However, the low selectivity of dehydrogenation reaction and high cost of commercial catalysts restrict the industrial applications. In this work, SiO_2, TiO_2 and Al_2O_3 are selected to synthesize Pd and Pt-based catalysts. For the Pd-based catalysts, we find that the dehydrogenation activity and selectivity are enhanced significantly by doping Cu, Ni and Co due to the bimetallic synergistic effect. Pt/TiO_2 and Pt/Al_2O_3 exhibit higher catalytic activity and selectivity than the Pd-based catalysts with similar metal load. The calculated TOFs are partially dependent on the particle size and dispersity of catalysts, suggesting that the dehydrogenation process is structure sensitive. After optimization for the Pt/TiO_2 catalysts, 1.0 wt% Pt/TiO_2 shows the 99.3% dehydrogenation capacity at 453 K, evidently higher than those of commercial Pd and Pt-based catalysts. Combined with the analysis of XRD, XPS, TEM, TPR, it is found that the enhancement of 12H-NECZ dehydrogenation efficiency is attributed to the promotion on the activation of the intermediates due to the strong metal-support interaction (SMSI) and surface structure of catalysts. This research can help reveal the hydrogen storage mechanism and provide guidance for designing new catalysts.

Fluorinated Nanoporous Organic Polymers for Micro-Organic Pollutants Removal from Water

ENGINEERING
SUSTAINABLE
DEVELOPMENT

Conference: Technical and Engineering Challenges of Addressing Sustainable Development

Year: 2019

Proceeding: Engineering Sustainable Development 2019

Group: Poster Submissions

Session: Poster Session

Authors:

Kim, D., KAIST

Nguyen, T. S., KAIST
Yavuz, C. T., UCSB
Rozyyev, V., KAIST
Lee, J., KAIST

Design of nanoporous polymers that contain confined spaces with functionalities can lead to selective chemical interactions and separation of species with similar sizes and functionality. Micro-organic pollutants are emerging contaminants such as pharmaceutics, endocrine disrupting chemicals (EDCs), and pesticides, although widely available activated carbon and salt-based flocculants operate sufficiently for conventional water treatment procedures. We recently reported a charge-specific size-dependent separation of water-soluble molecules through an ultra-microporous polymeric network that features fluorines as the predominant surface functional groups (Nature Commun., 7, 13377, 2016). Treatment of similarly sized organic molecules with and without charges shows that fluorine interacts with charges favorably. Control experiments using similarly constructed frameworks with or without fluorines verify the fluorine-cation interactions. The fluorinated nano-porous organic polymers can contribute to the removal of a wide range of micropollutants during water and wastewater treatment. Similarly, applying these concepts to conventional adsorbents like activated carbon opens new paths for sustainable water treatment.

AUTHOR INDEX

AUTHOR INDEX